Archimedes, the Center of Gravity, and the First Law of Mechanics

Andre K.T. Assis

Apeiron
Montreal

Published by C. Roy Keys Inc.
4405, rue St-Dominique
Montreal, Quebec H2W 2B2 Canada
http://redshift.vif.com

First Published 2008

Library and Archives Canada Cataloguing in Publication

Assis, André Koch Torres, 1962-
Archimedes, the center of gravity, and the first law of mechanics / Andre K.T. Assis.

Includes bibliographical references.
ISBN 978-0-9732911-6-2

1. Center of mass--Textbooks. 2. Center of mass--Experiments. 3. Mechanics--Textbooks. 4. Mechanics--Experiments. I. Title.

QA839.A87 2008 531'.14 C2007-907265-8

Front cover: Oil painting of Archimedes by the Italian artist Domenico Fetti (1589-1624), painted about 1620.

Back cover: Photos of a few of the experiments described in this book. A horizontal pasteboard triangle supported at the barycenter by a vertical stick. A rectangle and a plumb line suspended by a needle. An equilibrist upside down supported at the head, with modeling clay on his hands. A lever in equilibrium with different weights on each arm.

Table of Contents

Introduction

One of the goals of this book is to present the basic phenomena of mechanics through simple experiments performed with inexpensive materials. We present the fundamental experiments on falling bodies, equilibrium and oscillations around equilibrium positions. We also show how the theoretical concepts are formed and modified during this process, just as occurred in the formulation of the basic laws of mechanics.

We show how more complex phenomena can be explained and clarified by means of elementary experiments. Playful and curious experiments are also presented. They stimulate creativity, critical thinking and a sense of humour in science. They also relate everyday phenomena to the fundamental laws of physics.

The emphasis is placed on experimental activities. After the experiments we formulate the definitions, concepts, postulates, principles, and laws describing the phenomena. The materials utilized are very simple, easily found at home or in stores, all of them very inexpensive. Even so, we can carry out very precise experiments and construct sensitive scientific equipment. The reader need not depend on any school or research laboratory, as he can build his own equipment and perform all the measurements.

If the experiments presented here are performed in the classroom, each student should ideally perform all the tasks, even when working in a group. Each one should build his own equipment (support, plumb line, lever, etc.), cut out his geometric figures and then take all this personal material home. This procedure is richer in lessons than simple demonstrations of the effects by a teacher. It is essential that all students put their hands to the plough.

The book is also rich in historical information, which gives the context in which some laws were discovered, and also different approaches taken in discovering them. We are careful about in formulating concepts and physical principles. It will be seen, for example, how difficult is to find the correct words to precisely define the center of gravity so that this concept can encompass a whole series of experiments. We distinguish clearly between definitions, postulates, experimental results, and physical laws. We also distinguish explanations from descriptions of phenomena. These aspects illustrate the sociological and human aspects of the formulation of physical laws.

This book is written for students and teachers of science, physics, and mathematics. It can be utilized at High Schools or at Universities, depending on the level at which each aspect is analyzed and explored. It has enough experimental and theoretical material to be employed in all levels of teaching. Each teacher should adapt the contents presented here to his own school environment. It can also be utilized in courses on the history and philosophy of science.

The best way to grasp the contents of the book is to perform the majority of the experiments described here in parallel with the reading. There are many philosophical, theoretical, and mathematical approaches relating to physical science. But physics is essentially an experimental science. It is the combination of all these aspects that make it so fascinating. For this reason we strongly recommend that the experiments presented in the book be repeated and improved. We hope that the reader will have the same pleasure in performing these experiments as we had in developing them.

When necessary we employ the sign $\equiv$ as a symbol of definition. We utilize the SI international system of units.

To all those who, down through the centuries, have worked to preserve, translate, interpret, and disseminate the works of Archimedes.

Acknowledgments

The motivation to write this book arose from courses we gave to high school science teachers over the past few years. The exchange of ideas with these teachers and with our collaborators at the University were very rich and stimulating.

The inspiration for the majority of the experiments on equilibrium and the center of gravity (CG) of bodies came from the excellent works of Norberto Ferreira and Alberto Gaspar.[1]

We also thank for suggestions, references and ideas: Norberto Ferreira, Alberto Gaspar, Rui Vieira, Emerson, Dicesar Lass Fernandez, Silvio Seno Chibeni, César José Calderon Filho, Pedro Leopoldo e Silva Lopes, Fábio Miguel de Matos Ravanelli, Juliano Camillo, Lucas Angioni, J. Len Berggren, Henry Mendell, and Roy Keys, as well as our students at the Institute of Physics with whom we discussed these ideas. Steve Hutcheon read the whole manuscript and suggested many improvements. My daughter and Eduardo Meirelles helped with the figures.

Special thanks are due to the Institute of Physics, to the Institute of Mathematics, to the Grupo Gestor de Projetos Educacionais (GGPE) and to FAEPEX of the State University of Campinas, in Brazil, which provided the necessary conditions for the preparation of this essay.

Andre K.T. Assis

Institute of Physics, State University of Campinas

13083-970 Campinas-SP, Brazil

E-mail: assis@ifi.unicamp.br

homepage: http://www.ifi.unicamp.br/~assis/

[1] N. Ferreira, *Mecânica*, Projeto RIPE – Rede de Instrumentação para o Ensino (São Paulo: Instituto de Física, USP.); N. Ferreira, *Equilíbrio* (São Paulo: Instituto de Física, USP, 2007), http://www.ludoteca.if.usp.br/; A. Gaspar, *Experiências de Ciências para o Ensino Fundamental* (São Paulo: Ática, 2003).

Chapter 1

The Life of Archimedes

The account of Archimedes's life given here is drawn essentially from Heath,[1] Dijksterhuis,[2] and Plutarch.[3]

Archimedes lived from 287 to 212 B.C. He was born in Syracuse, on the coast of Sicily, where he spent most of his life. He was the son of Pheidias, an astronomer, who estimated the ratio of the diameters of the Sun and the Moon. Archimedes spent some time in Egypt. It is possible that he studied at the city of Alexandria, which was then the center of Greek science, with the successors of the mathematician Euclid, who flourished around 300 B.C. and published the book *The Elements of Geometry*. Many of Archimedes's works were sent to mathematicians who lived in Alexandria or who had been there. The famous Museum in Alexandria, which housed a huge library, one of the largest in antiquity, was founded around 300 B.C. It is estimated that it had up to 500,000 papyrus scrolls, with an average of 20,000 words in each scroll. The city was under Roman rule from 30 B.C. to 400 A.D. When Cesar was besieged in the palace of Alexandria in 48 B.C., a fire may have reached the book repository, and in 391 A.D. the library may have been destroyed by decree of Emperor Theodosius I. There are no records of the existence of the library and museum after the fifth century. The Roman Empire was fragmented into two parts, Western and Eastern, in 395. Many works of Archimedes were irremediably lost in the ensuing period.

Archimedes is considered one of the greatest scientists of all time, and the greatest mathematician of antiquity. In modern times only Isaac

[1] T.L. Heath, ed., *The Works of Archimedes* (Dover, New York, 2002). T.L. Heath, *A History of Greek Mathematics*, Vol. II: *From Aristarchus to Diophantus* (Oxford: Clarendon Press, 1921).

[2] E.J. Dijksterhuis, *Archimedes* (Princeton: Princeton University Press, 1987).

[3] Plutarch, *Marcellus*, http://classics.mit.edu/Plutarch/marcellu.html, translated by John Dryden, site available in 2007.

Newton (1642-1727) is comparable to him, both for producing experimental and theoretical works of great impact, and for his originality and immense influence. By utilizing the method of exhaustion, Archimedes was able to determine the area, volume, and center of gravity of many important geometrical figures, which had never been accomplished before him. He is considered one of the founders of statics and hydrostatics.

His concentration is well described in this passage from Plutarch (46-*c.* 122):[4]

> And thus it ceases to be incredible that (as is commonly told of him) the charm of his familiar and domestic Siren made him forget his food and neglect his person, to that degree that when he was occasionally carried by absolute violence to bathe or have his body anointed, he used to trace geometrical figures in the ashes of the fire, and diagrams in the oil on his body, being in a state of entire preoccupation, and, in the truest sense, divine possession with his love and delight in science.

Archimedes's preoccupation with scientific matters in all aspects of life is also recounted by Vitruvius (*c.* 90-20 B.C.) in a famous passage in his book on architecture. It is related to the fundamental principle of hydrostatics, which deals with the upward force exerted upon bodies immersed in fluids. The passage illustrates how Archimedes arrived at this principle, or at least the origin of the initial intuition which led to the discovery. We quote from Mach:[5]

> Though Archimedes discovered many curious matters that evince great intelligence, that which I am about to mention is the most extraordinary. Hiero, when he obtained the regal power in Syracuse, having, on the fortunate turn of his affairs, decreed a votive crown of gold to be placed in a certain temple to the immortal gods, commanded it to be made of great value, and assigned for this purpose an appropriate weight of the metal to the manufacturer. The latter, in due time, presented the work to the king, beautifully wrought; and the weight appeared to correspond with that of the gold which had been assigned for it. But a report had been circulated, that some of the gold had been abstracted, and that the deficiency thus caused had been supplied by silver, Hiero was indignant at the fraud, and, unacquainted with the method by which the theft might be detected, requested Archimedes would undertake to give it his attention. Charged with this commission, he by chance went to a bath, and on jumping into the tub, perceived that, just in the proportion that his body became immersed, in the same proportion the water ran out of the vessel.

[4] Plutarch, *op. cit.*

[5] E. Mach, *The Science of Mechanics* (La Salle: Open Court, 1960), 107.

> Whence, catching at the method to be adopted for the solution of the proposition, he immediately followed it up, leapt out of the vessel in joy, and returning home naked, cried out with a loud voice that he had found that of which he was in search, for he continued exclaiming, in Greek, εὕρηχα, εὕρηχα, (I have found it, I have found it!)

Those works of Archimedes that have survived were addressed to the astronomer Conon of Samos (at that time living in Alexandria), to Conon's disciple Dositheus after the death of Conon, to king Gelon, son of the king Hiero of Syracuse, and to Eratosthenes, librarian of the Library of Alexandria and famous for his precise estimation of the radius of the Earth.

Archimedes would send his works together with some introductory texts. Through these texts we can discover the order of some of his discoveries and a little of his personality. For example, in the introduction of his famous work *The Method*, he states:[6]

> Archimedes to Eratosthenes greeting.
>
> I sent you on a former occasion some of the theorems discovered by me, merely writing out the enunciation and inviting you to discover the proofs, which at the moment I did not give. The enunciations of the theorems which I sent were as follows. (...) The proofs then of these theorems I have written in this book and now send to you. (...)

His habit of sending initially only the enunciations of some theorems, without demonstrations, may have led some mathematicians plagiarize Archimedes, claiming that his results belonged to them. It is perhaps for this reason that Archimedes on one occasion sent two false results, as he mentions in the preface of his work *On Spirals*:[7]

> Archimedes to Dositheus greeting.
>
> Of most of the theorems which I sent to Conon, and of which you ask me from time to time to send you the proofs, the demonstrations are already before you in the books brought to you by Heracleides; and some more are contained in that which I now send you. Do not be surprised at my taking a considerable time before publishing the proofs. This has been owing to my desire to communicate them first to persons engaged in mathematical studies and anxious to investigate them. In fact, how many theorems in geometry which have seemed at first impracticable are in time successfully worked out! Now Conon died before he had sufficient time to investigate the theorems referred to; otherwise

[6] T.L. Heath, ed., *The Works of Archimedes* (New York: Dover, 2002), Supplement, pp. 12-13.
[7] *Ibid.*, p. 151.

> he would have discovered and made manifest all these things, and would have enriched geometry by many other discoveries besides. For I know well that it was no common ability that he brought to bear on mathematics, and that his industry was extraordinary. But, though many years have elapsed since Conon's death, I do not find that any one of the problems has been stirred by a single person. I wish now to put them in review one by one, particularly as it happens that there are two included among them which are impossible of realisation[8] [and which may serve as a warning] how those who claim to discover everything but produce no proofs of the same may be confuted as having actually pretended to discover the impossible.

Archimedes would often spend years trying to find the proof of a difficult theorem. We can see the perseverance with which he strived to reach his goal in the introduction to *On Conoids and Spheroids*:[9]

> Archimedes to Dositheus greeting.
>
> In this book I have set forth and send you the proofs of the remaining theorems not included in what I sent you before, and also of some others discovered later which, though I had often tried to investigate them previously, I had failed to arrive at because I found their discovery attended with some difficulty. And this is why even the propositions themselves were not published with the rest. But afterwards, when I had studied them with greater care, I discovered what I had failed in before.

Although the works that have come down to us are related to mathematics and theoretical physics, the fame of Archimedes in antiquity is due to his work as an engineer and builder of war machines (catapults, burning mirrors, *etc.*). One of the inventions attributed to him is a water pumping system known as the cochlias, or Archimedes screw, which is used even to this day. The word cochlias is Greek, meaning snail. It is believed that he invented this hydraulic machine during his stay in Egypt, where it was used for irrigating fields and pumping water.

He built a famous planetarium that had a single hydraulic mechanism which moved several globes simultaneously, reproducing the motions of the stars, the Sun, the Moon, and the planets around the Earth. He also built a hydraulic organ in which the air fed to the pipes was compressed above water in an air chamber. Also attributed to him are the inventions of the compound pulley, machines for discharging showers of missiles, the Roman balance with unequal arms *etc.*

---

[8] Note by Heath: Heiberg reads *τέλος δέ ποθεσόμενα*, but F has τέλους, so that the true reading is perhaps τέλους δέ ποτιδεόμενα. The meaning appears to be simply 'wrong.'

[9] Heath, ed., *The Works of Archimedes*, p. 99.

Several authors quote a famous sentence by Archimedes in connection with his mechanical devices and his ability to move great weights with a small force: "Give me a place to stand on, and I will move the Earth."[10] Apparently he uttered this when he accomplished a feat ordered by king Hiero to launch a ship weighing many tons and carrying a full load. He succeeded in this unaided, with his hands and the aid of a few mechanical instruments. Plutarch relates this story as follows:[11]

> Archimedes, however, in writing to King Hiero, whose friend and near relation he was, had stated that given the force, any given weight might be moved, and even boasted, we are told, relying on the strength of demonstration, that if there were another earth, by going into it he could remove this. Hiero being struck with amazement at this, and entreating him to make good this problem by actual experiment, and show some great weight moved by a small engine, he fixed accordingly upon a ship of burden out of the king's arsenal, which could not be drawn out of the dock without great labour and many men; and, loading her with many passengers and a full freight, sitting himself the while far off, with no great endeavour, but only holding the head of the pulley in his hand and drawing the cords by degrees, he drew the ship in a straight line, as smoothly and evenly as if she had been in the sea.

Hiero was so amazed that he said: "From that day forth Archimedes was to be believed in everything that he might say."[12]

Plutarch continued:[13]

> The king, astonished at this, and convinced of the power of the art, prevailed upon Archimedes to make him engines accommodated to all the purposes, offensive and defensive, of a siege. These the king himself never made use of, because he spent almost all his life in a profound quiet and the highest affluence. But the apparatus was, in most opportune time, ready at hand for the Syracusans, and with it also the engineer himself.

During the second Punic war between Rome and Carthage, the city of Syracuse was allied with Carthage. Syracuse was attacked by the Romans in 214 B.C., under General Marcellus. Many histories about Archimedes have survived in a famous biography of Marcellus written by Plutarch. Marcellus attacked Syracuse by land and sea, heavily armed. According to Plutarch:[14]

---

[10] E.J. Dijksterhuis, *Archimedes* (Princeton: Princeton University Press, 1987), p. 15.

[11] Plutarch, *Marcellus*.

[12] Heath, ed., *The Works of Archimedes*, p. xix

[13] Plutarch, *Marcellus*.

[14] Plutarch, *Marcellus*.

> [All machines of Marcellus], however, were, it would seem, but trifles for Archimedes and his machines. These machines he had designed and contrived, not as matters of any importance, but as mere amusements in geometry; in compliance with King Hiero's desire and request, some little time before, that he should reduce to practice some part of his admirable speculation in science, and by accommodating the theoretic truth to sensation and ordinary use, bring it more within the appreciation of the people in general.

Elsewhere, Plutarch writes:[15]

> When, therefore, the Romans assaulted the walls in two places at once, fear and consternation stupefied the Syracusans, believing that nothing was able to resist that violence and those forces. But when Archimedes began to ply his engines, he at once shot against the land forces all sorts of missile weapons, and immense masses of stone that came down with incredible noise and violence; against which no man could stand; for they knocked down those upon whom they fell in heaps, breaking all their ranks and files. In the meantime huge poles thrust out from the walls over the ships sunk some by the great weights which they let down from on high upon them; others they lifted up into the air by an iron hand or beak like a crane's beak and, when they had drawn them up by the prow, and set them on end upon the poop, they plunged them to the bottom of the sea; or else the ships, drawn by engines within, and whirled about, were dashed against steep rocks that stood jutting out under the walls, with great destruction of the soldiers that were aboard them. (...) In fine, when such terror had seized upon the Romans, that, if they did but see a little rope or a piece of wood from the wall, instantly crying out, that there it was again, Archimedes was about to let fly some engine at them, they turned their backs and fled, Marcellus desisted from conflicts and assaults, putting all his hope in a long siege.

Also connected with the defence of Syracuse is the famous story about burning the Roman ships with mirrors. Archimedes used a great mirror or a system of small mirrors in order to concentrate the sun's rays and focus them on the ships. The two most famous accounts are due to Johannes Tzetzes, a Byzantine scholar, and John Zonaras, both of the twelfth century:

> When Marcellus withdrew them [his ships] a bow-shot, the old man [Archimedes] constructed a kind of hexagonal mirror, and at an interval proportionate to the size of the mirror he set similar small mirrors with four edges, moved by links and by a form of hinge, and made it the centre of the sun's beams--its noon-tide

[15] Plutarch, *Marcellus.*

> beam, whether in summer or in mid-winter. Afterwards, when the beams were reflected in the mirror, a fearful kindling of fire was raised in the ships, and at the distance of a bow-shot he turned them into ashes. In this way did the old man prevail over Marcellus with his weapons. –J. Tzetzes.[16]

> At last in an incredible manner he [Archimedes] burned up the whole Roman fleet. For by tilting a kind of mirror toward the sun he concentrated the sun's beam upon it; and owing to the thickness and smoothness of the mirror he ignited the air from this beam and kindled a great flame, the whole of which he directed upon the ships that lay at anchor in the path of the fire, until he consumed them all. –J. Zonaras.[17]

Only after a siege of three years was Marcellus able to conquer Syracuse. Archimedes was killed by a Roman soldier in 212 A.D. during the capture of the city. Marcellus had given express orders that Archimedes's life should be spared, in recognition of the genius of this enemy who had caused him so many losses. In spite of this, a soldier killed him while he was trying to protect or finish some mathematical discoveries. The last words uttered by Archimedes seem to have been addressed to this soldier: "Fellow, stand away from my diagram."[18] Plutarch gives us three different versions of his death:[19]

> But nothing afflicted Marcellus so much as the death of Archimedes, who was then, as fate would have it, intent upon working out some problem by a diagram, and having fixed his mind alike and his eyes upon the subject of his speculation, he never noticed the incursion of the Romans, nor that the city was taken. In this transport of study and contemplation, a soldier, unexpectedly coming up to him, commanded him to follow to Marcellus; which he declining to do before he had worked out his problem to a demonstration, the soldier, enraged, drew his sword and ran him through. Others write that a Roman soldier, running upon him with a drawn sword, offered to kill him; and that Archimedes, looking back, earnestly besought him to hold his hand a little while, that he might not leave what he was then at work upon inconclusive and imperfect; but the soldier, nothing moved by his entreaty, instantly killed him. Others again relate that, as Archimedes was carrying to Marcellus mathematical instruments, dials, spheres, and angles, by which the magnitude of the sun might be measured to the sight, some soldiers seeing him, and thinking that he carried gold in a vessel, slew him. Certain it

[16] C. Rorres (organizer), *Archimedes*, http://www.math.nyu.edu/~crorres/Archimedes/contents.html. Site available in 2007.

[17] Rorres, *Archimedes*.

[18] Dijksterhuis, *Archimedes*, p. 31.

[19] Plutarch, *Marcellus*.

> is that his death was very afflicting to Marcellus; and that Marcellus ever after regarded him that killed him as a murderer; and that he sought for his kindred and honoured them with signal favours.

During his lifetime, Archimedes expressed the wish that upon his tomb there should be placed a cylinder circumscribing a sphere within it, together with an inscription giving the ratio between the volumes of these two bodies. We can infer that he regarded the discovery of this ratio as his greatest achievement. It appears in Propositions 33 and 34 of the first part of his work *On the Sphere and Cylinder*. These two results are extremely important, and both are due to Archimedes. "Proposition 33: The surface of any sphere is equal to four times the greatest circle in it."[20] That is, in modern language, with A being the area of the surface of a sphere of radius r: $A = 4(\pi r^2)$. "Proposition 34: Any sphere is equal to four times the cone which has its base equal to the greatest circle in the sphere and its height equal to the radius of the sphere."[21] In modern language, with V_E the volume of a sphere of radius r, and $V_C = \pi r^2 \cdot r/3$, the volume of a cone of height r and base area equal to πr^2, we have $V_E = 4V_C = 4(\pi r^3/3)$. The inscription Archimedes requested for his tomb seems to be related to a Corollary presented at the end of this Proposition: "From what has been proved it follows that every cylinder whose base is the greatest circle in a sphere whose height is equal to the diameter of the sphere is a $3/2$ of the sphere, and its surface together with its base is $3/2$ of the surface of the sphere."[22]

Marcellus saw to it that this wish was fulfilled. Cicero (106-43 B.C.), the Roman orator, saw this tomb in a neglected state in 75 A.D. when he was quaestor in Sicily, and restored it. It has never been seen since. Cicero' writes:[23]

> But from Dionysius's own city of Syracuse I will summon up from the dust—where his measuring rod once traced its lines—an obscure little man who lived many years later, Archimedes. When I was quaestor in Sicily I managed to track down his grave. The Syracusians knew nothing about it, and indeed denied that any such thing existed. But there it was, completely surrounded and hidden by bushes of brambles and thorns. I remembered having heard of some simple lines of verse which had been inscribed on his tomb, referring to a sphere and cylinder modelled in stone on top of the grave. And so I took a good look round all the numerous tombs that stand beside the Agrigentine

[20] Heath, ed., *The Works of Archimedes*, p. 39.
[21] Heath, ed., *The Works of Archimedes*, p. 41
[22] Heath, ed., *The Works of Archimedes*, p. 43.
[23] Rorres, *Archimedes*.

Gate. Finally I noted a little column just visible above the scrub: it was surmounted by a sphere and a cylinder. I immediately said to the Syracusans, some of whose leading citizens were with me at the time, that I believed this was the very object I had been looking for. Men were sent in with sickles to clear the site, and when a path to the monument had been opened we walked right up to it. And the verses were still visible, though approximately the second half of each line had been worn away. So one of the most famous cities in the Greek world, and in former days a great centre of learning as well, would have remained in total ignorance of the tomb of the most brilliant citizen it had ever produced, had a man from Arpinum not come and pointed it out!

Chapter 2

The Works of Archimedes

The works of Archimedes known to us are available in the original Greek and in Latin.[1] English translations in modern notation have been published.[2] A literal translation from the Greek to French can also be found.[3]

Until one hundred years ago, the oldest and most important manuscripts containing works of Archimedes in Greek (with the exception of *The Method*, which did not appear in any manuscript) were mainly from the 15th and 16th centuries, housed in libraries located in Europe. They had been copied from two other 9th and 10th century Greek manuscripts. One of these manuscripts belonged to the humanist George Valla, who taught at Venice between 1489 and 1499. This manuscript disappeared between 1544 and 1564. It is not known if it still exists. It contained the following works, in this order: *On the Sphere and Cylinder*, *Measurement of a Circle*, *On Conoids and Spheroids*, *On Spirals*, *On the Equilibrium of Planes*, *The Sand-Reckoner*, *Quadrature of the Parabola*, *Eutocius's commentaries on the Sphere and Cylinder*, *Measurement of a Circle*, and *On the Equilibrium of Planes*.

The last record of the second of the 9th and 10th century manuscripts was in the Vatican Library in the years 1295 and 1311. It is not known if this manuscript still exists. It contained the following works, in this order: *On Spirals*, *On the Equilibrium of Planes*, *Quadrature of the Parabola*, *Measurement of a Circle*, *On the Sphere and Cylinder*, *Eutocius's commentaries on the Sphere and Cylinder*, *On Conoids and Spheroids*, *Eutocius's commentaries on the Equilibrium of Planes*, *On Float-*

[1] J.L. Heiberg, *Archimedis Opera*. Second edition, in 3 volumes (Leipzig: Teubner, 1910-1915).

[2] T.L. Heath, ed., *The Works of Archimedes* (New York: Dover, 2002), E.J. Dijksterhuis, *Archimedes* (Princeton: Princeton University Press, 1987).

[3] C. Mugler, *Les Oeuvres d'Archimède* (Paris: Budé, 1970-2). Vol. 1-4.

ing Bodies. The latter work on floating bodies, in two parts, was not contained in the first manuscript.

The work *On Floating Bodies* was only known until 1906 from a Latin translation made by the Flemish Dominican Willem van Moerbeke in 1269, based on the second 9th or 10th century manuscript. He translated all of Archimedes's works to which he had access into Latin, and this was very important in spreading of Archimedes's ideas. The original manuscript containing Moerbeke's translation was found again in Rome in 1884, and is now at the Vatican Library.

Archimedes wrote in the Doric dialect. In the manuscripts still extant his original language was transformed in some books totally, in others only partially, into the Attic dialect common in Greece. In the 9th century some of his works were translated to Arabic. The first Latin translations of the works of Archimedes and of several scientists and philosophers of Greece were made during the 12th and 13th centuries. Gutenberg invented movable type for the printing press in Europe in the mid-15th century. The publication of Archimedes's works in printed form began in the 16th century, the oldest being from 1503, containing the *Measurement of a Circle* and the *Quadrature of the Parabola*. Printed in 1544, the *editio princeps* contained the major known works by Archimedes, in Greek and Latin, with the exception of *On Floating Bodies*. The invention of the press was very important for the spread of his ideas. The first translations of some of his works to a living language, German, were published in 1667 and 1670, by J.C. Sturm. In 1807 the first French translation of all his known works was made by Peyrard. In 1897 and 1912 the first English translation was published by Sir T.L. Heath.

2.1 Extant works

We present here the extant works of Archimedes in the order in which they were written according to Heath.[4] Much controversy surrounds this chronology. Knorr, for example, places *The Method* at the end of his works.[5]

[4] T.L. Heath, *A History of Greek Mathematics*, Vol. II: *From Aristarchus to Diophantus* (Oxford: Clarendon Press, 1921), pp. 22-23.

[5] W.R. Knorr, "Archimedes and the Elements: Proposal for a Revised Chronological Ordering of the Archimedean Corpus," *Archive for the History of Exact Sciences* **19** (1978/79): 211-290

***On the Equilibrium of Planes*, or**
***The Centers of Gravity of Planes.* Book I**

Archimedes derives the law of the lever theoretically utilizing the axiomatic method and calculates the center of gravity of parallelograms, triangles, and trapeziums.

Quadrature of the Parabola

Archimedes finds the area of a parabolic segment. Proposition 24: "Every segment bounded by a parabola and a chord Qq is equal to four-thirds of the triangle which has the same base as the segment and equal height."[6] He presents two proofs of this result. In the first, he performs a mechanical quadrature, utilizing the law of the lever. In the second, he performs a geometric quadrature.

***On the Equilibrium of Planes*, or**
***The Centers of Gravity of Planes.* Book II**

Archimedes finds the center of gravity of a parabolic segment.

***The Method of Treating Mechanical Problems*, to Eratosthenes**

This work is usually called simply *The Method*. Archimedes presents a mechanical method to obtain geometrical results (calculation of areas, volumes and centers of gravity) utilizing the law of the lever and concepts of the theory of the center of gravity. He presents several examples of this heuristic method which he created and employed, illustrating how to apply it. He thus obtains the quadrature of the parabola, the volume and center of gravity of any segment of a sphere, the center of gravity of a semi-circle, the center of gravity of a paraboloid of revolution, and several other results. This work will be discussed in more detail in a later Section.

***On the Sphere and Cylinder*, Books I and II**

Archimedes shows that the area of the surface of a sphere is equal to four times the greatest circle passing through the center of the sphere; finds the area of any segment of the sphere; shows that the volume of the sphere is equal to two-thirds the volume of the circumscribed cylinder, and that the surface of the sphere is equal to two-thirds the surface of the circumscribed cylinder, including the bases. In the second part of this work, the most important result is how to divide a sphere by a plane in such a way that the ratio of the volumes of the two segments has a given value.

---

[6] Heath, ed., *The Works of Archimedes*, p. 251.

On Spirals

Archimedes defines a spiral through the uniform motion of a point along a straight line, this straight line rotating with a constant angular velocity in a plane. He establishes the fundamental properties of the spiral, relating the length of the radius vector to the angles of revolution that generate the spiral. He presents results related to the tangents of the spiral, and shows how to calculate areas of parts of the spiral.

As a curiosity we quote here the first two propositions and the main definition presented in this work. This spiral is represented nowadays in polar coordinates by the relation $\rho = k\varphi$, where k is a constant, ρ is the distance to the z-axis (or from the origin, considering the motion in the xy plane) and φ is the angle of the radius vector relative to the x axis. In this representation the time does not appear. On the other hand, the historical relevance of the original definition given by Archimedes is the introduction of the time concept in geometry. This was crucial for the later development of classical mechanics.

> Proposition 1: If a point move at a uniform rate along any line, and two lengths be taken on it, they will be proportional to the times of describing them.[7]
>
> Proposition 2: If each of two points on different lines respectively move along them each at a uniform rate, and if lengths be taken, one on each line, forming pairs, such that each pair are described in equal times, the lengths will be proportionals.[8]
>
> Definition: If a straight line drawn in a plane revolve at a uniform rate about one extremity which remains fixed and return to the position from which it started, and if, at the same time as the line revolves, a point move at a uniform rate along the straight line beginning from the extremity which remains fixed, the point will describe a spiral in the plane.[9]

On Conoids and Spheroids

Archimedes studies the paraboloids of revolution, the hyperboloids of revolution (conoids) and the ellipsoids (spheroids) obtained by the rotation of an ellipse around one of its axes. The main goal of the work is to investigate the volume of segments of these three-dimensional bodies. He shows, for example, in Propositions 21 and 22, that the volume of a paraboloid of revolution is $3/2$ of the volume of the cone which has the same base and the same height. Analogous, but more complex results, are obtained for the hyperboloid of revolution and for the ellipsoid.

---

[7] Heath, ed., *The Works of Archimedes*, p. 155.

[8] *Ibid.*

[9] Ibid, p. 165.

***On Floating Bodies.* Books I and II**

Archimedes establishes the fundamental principles of hydrostatics, giving the weight of a body immersed in a fluid. He also studies the conditions of stability of a spherical segment floating in a fluid, and of a paraboloid of revolution floating in a fluid.

In the first part of this work, Archimedes creates the entire science of hydrostatics. We know of no other author who worked with this subject prior to him. His basic postulate reads as follows:[10]

> Postulate: Let it be granted that the fluid is of such a nature that of the parts of it which are at the same level and adjacent to one another that which is pressed the less is pushed away by that which is pressed the more, and that each of its parts is pressed by the fluid which is vertically above it, if the fluid is not shut up in anything and is not compressed by anything else.

Heath's translation of this postulate reads as follows:

> Postulate 1: Let it be supposed that a fluid is of such a character that, its parts lying evenly and being continuous, that part which is thrust the less is driven along by that which is thrust the more; and that each of its parts is thrust by the fluid which is above it in a perpendicular direction if the fluid be sunk in anything and compressed by anything else.[11]

Heath's translation, published in 1897, was based on the Latin translation by Moerbeke in 1269, as the original text by Archimedes in Greek had been lost. In 1906, Heiberg found another manuscript containing the original Greek text of this work. Some parts of this manuscript remain undecipherable, and others are missing. In any event it contains this basic postulate, which clarifies the meaning of the last passage. Instead of Heath's "and that each of its parts is thrust by the fluid which is above it in a perpendicular direction if the fluid be sunk in anything and compressed by anything else," the correct meaning is that of Dijksterhuis or Mugler, namely, "that each of its parts is pressed by the fluid which is vertically above it, if the fluid is not shut up in anything and is not compressed by anything else."

Beginning with this postulate he arrives at an explanation for the spherical shape of the Earth, supposing it to be wholly composed of water. Then he proves the fundamental principle of hydrostatics, known today as Archimedes's principle, in Propositions 5 to 7. When he says that a solid is heavier or lighter than a fluid, he is referring to the relative or specific weight, that is, if the solid is more or less dense than a fluid.

[10] Dijksterhuis, *Archimedes*, p. 373; Mugler, *Les Oeuvres d'Archimède*, Vol. 3, p. 6.
[11] Heath, ed., *The Works of Archimedes*, p. 253.

> Proposition 5: Any solid lighter than a fluid will, if placed in the fluid, be so far immersed that the weight of the solid will be equal to the weight of the fluid displaced.[12]
>
> Proposition 6: If a solid lighter than a fluid be forcibly immersed in it, the solid will be driven upwards by a force equal to the difference between its weight and the weight of the fluid displaced.[13]
>
> Proposition 7: A solid heavier than a fluid will, if placed in it, descend to the bottom of the fluid, and the solid will, when weighed in the fluid, be lighter than its true weight by the weight of the fluid displaced.[14]

Based on these propositions at the end of the first book he determines the equilibrium conditions of a spherical segment floating in a fluid. In the second part, Archimedes presents a complete investigation of the conditions of equilibrium of a segment of a paraboloid of revolution floating in a fluid. His interest here seems very clear, namely, to study theoretically the stability of ships, although this is not explicitly mentioned. This is a work of applied mathematics, or theoretical engineering.

This is a monumental work which, for some two thousand years, was almost the only text on this topic. It was revived in the renaissance, influencing the works of Stevin (1548-1620) and Galileo (1564-1642).

Measurement of a Circle

This work does not come down to us in its original form. It is probably only a fragment of a larger text. Archimedes shows that the area of a circle is equal to the area of a right-angled triangle whose legs are the radius of the circle and the rectified circumference: "Proposition 1: The area of any circle is equal to a right-angled triangle in which one of the sides about the right angle is equal to the radius, and the other to the circumference, of the circle."[15] He also shows that the exact value of π is between $3\,^{10}/_{71} = 3.1408$ and $3\,^{1}/_{7} = 3.1429$. This he obtained by circumscribing and inscribing a circle with regular polygons of 96 sides. In his own words, "Proposition 3: The ratio of the circumference of any circle to its diameter is less than $3\,^{1}/_{7}$ but greater than $3\,^{10}/_{71}$."[16] In the middle of this proposition he presents precise approximations for the square roots of many numbers, without specifying how he arrived at these results. He

[12] Heath, ed., *The Works of Archimedes*, p. 257.
[13] *Ibid.*
[14] Heath, ed., *The Works of Archimedes*, p. 258.
[15] Heath, ed., *The Works of Archimedes*, p. 91.
[16] Heath, ed., *The Works of Archimedes*, p. 93.

states, for example, that $\frac{265}{153} < \sqrt{3} < \frac{1351}{780}$, or, $1.7320261 < \sqrt{3} < 1.7320513$.

The Sand-Reckoner

Archimedes deals with the problem of counting the number of grains of sand contained in the sphere of the fixed stars, utilizing estimations by Eudoxus, his father Pheidias, and Aristarchus. He proposes a numerical system capable of expressing numbers equivalent to our 8×10^{63}. It is in this work that Archimedes mentioned that the addition of the orders of the numbers (the equivalent of their exponents when the base is 10^8) corresponds to finding the product of these numbers.[17] This is the principle that led to the invention of logarithms many centuries later.

Also in this work, Archimedes mentions the heliocentric system of Aristarchus of Samos (*c.* 310-230 B.C.). The work of Aristarchus describing his heliocentric system has not been preserved. Here we reproduce the introduction of the Sand-Reckoner. This introduction is the oldest and most important evidence concerning the existence of a heliocentric system in antiquity. Due to this extremely important idea, Aristarchus is often called the Copernicus of antiquity. At the end of the introduction, Archimedes refers to a work called *Principles*, which is probably the title of one of Archimedes's works containing a system of expressing numbers that had been sent to Zeuxippus, and is quoted in the introduction. This work is not extant. Archimedes writes:

> There are some, king Gelon, who think that the number of the sand is infinite in multitude; and I mean by the sand not only that which exists about Syracuse and the rest of Sicily but also that which is found in every region whether inhabited or uninhabited. Again there are some who, without regarding it as infinite, yet think that no number has been named which is great enough to exceed its multitude. And it is clear that they who hold this view, if they imagined a mass made up of sand in other respects as large as the mass of the earth, including in it all the seas and the hollows of the earth filled up to a height equal to that of the highest of the mountains, would be many times further still from recognising that any number could be expressed which exceeded the multitude of the sand so taken. But I will try to show you by means of geometrical proofs, which you will be able to follow, that, of the numbers named by me and given in the work which I sent to Zeuxippus, some exceed not only the number of the mass of sand equal in magnitude to the earth filled up in the way described, but also that of a mass equal in magnitude to the universe. Now you are aware that 'universe' is the name given by most astronomers to the sphere whose centre is the centre of the

[17] Dijksterhuis, *Archimedes*, pp. 360-373.

> earth and whose radius is equal to the straight line between the centre of the sun and the centre of the earth. This is the common account (τά γραφόμενα), as you have heard from astronomers. But Aristarchus of Samos brought out a book consisting of some hypotheses, in which the premises lead to the result that the universe is many times greater than that now so called. His hypotheses are that the fixed stars and the sun remain unmoved, that the earth revolves about the sun in the circumference of a circle, the sun lying in the middle of the orbit, and that the sphere of the fixed stars, situated about the same centre as the sun, is so great that the circle in which he supposes the earth to revolve bears such a proportion to the distance of the fixed stars as the centre of the sphere bears to its surface. Now it is easy to see that this is impossible; for, since the centre of the sphere has no magnitude, we cannot conceive it to bear any ratio whatever to the surface of the sphere. We must however take Aristarchus to mean this: since we conceive the earth to be, as it were, the centre of the universe, the ratio which the earth bears to what we describe as the 'universe' is the same as the ratio which the sphere containing the circle in which he supposes the earth to revolve bears to the sphere of the fixed stars. For he adapts the proofs of his results to a hypothesis of this kind, and in particular he appears to suppose the magnitude of the sphere in which he represents the earth as moving to be equal to what we call the 'universe.'
>
> I say then that, even if a sphere were made up of the sand, as great as Aristarchus supposes the sphere of the fixed stars to be, I shall still prove that, of the numbers named in the Principles, some exceed in multitude the number of the sand which is equal in magnitude to the sphere referred to, provided that the following assumptions be made(...)[18]

It is also known that Archimedes wrote other works which exist today only in fragments or in references by other writers:

The Cattle-Problem

This is contained in an epigram communicated by Archimedes to the mathematicians of Alexandria in a letter to Eratosthenes. It is a problem of algebra with 8 unknowns. The complete solution leads to a number with 206,545 digits.

Book of Lemmas

A collection of important lemmas relating to planimetric figures.

Semi-Regular Polyhedra

The regular polyhedra were known by Plato and are described by Euclid in his book *The Elements of Geometry*. Their faces are composed of

[18] Heath, ed., *The Works of Archimedes*, pp. 221-2.

regular equal polygons, equilateral and equiangular. There are only 5 regular platonic solids: the tetrahedron, the cube, the octahedron, the dodecahedron and the icosahedron.

In this work Archimedes describes the construction of the semi-regular polyhedra which he discovered. Its faces are regular polygons, but with different numbers of sides, such as squares and equilateral triangles. There are only 13 of these solids, all discovered by Archimedes. They are called Archimedian polyhedra.

The Stomachion

This is a game like tangram, with 14 pieces which together form a square. Archimedes probably tried to find the number of ways in which these 14 pieces can be put together in order to form a square.

Area of the Triangle

Some authors consider that Archimedes discovered the expression usually attributed to Heron in the first century A.D. for the area of a triangle in terms of its sides.

Construction of a Regular Heptagon

Archimedes presents the construction of a heptagon inscribed within a circle.

Other works mentioned by Archimedes or by other authors are not extant. In some cases we know only the title, or have a general idea of their content. The same work may be cited with different names:

Principles, or Naming of Numbers

On how to express large numbers.

On the Centers of Gravity

Elements of Mechanics

On the center of gravity and law of the lever. The work *On the Equilibrium of Planes* is probably only a small part of this larger work.

Equilibria

On the CG of solids.

Book on Columns*, or *Book of Supports

According to Heron, Archimedes dealt here with bodies supported by two or more columns. He solved the problem of finding which part of the total weight of the body was supported by each pillar.

On Balances*, or *On Levers
On the center of gravity and the law of the lever.

One work on optics
Including the law of reflection and studies on refraction.

On Sphere-Making
A mechanical work describing the construction of a sphere representing the motions of the celestial bodies, probably a description of the famous planetarium built by Archimedes.

On the Calendar
On the length of the year.

On Circles Touching One Another

On Parallel Lines

On Triangles

On Properties of Right-Angled Triangles

On the Assumptions for the Elements of Geometry

Book of Data or Definitions

2.2 *The Method*

Of all the Archimedes's works known today, the one that has received the greatest attention is *The Method*. One of the few things known about this work until 1906 was its title. Between 1880 and 1881 the Danish scholar J.L. Heiberg (1854-1928), a professor of classical philology at Copenhagen University, published the complete works of Archimedes then known, in Greek and Latin, in three volumes. This book was utilized as the basis for the modern translation of his works into many living languages, such as the English made by T. L. Heath (1861-1940) and published in 1897. When he described the lost works of Archimedes, Heath quoted *The Method* in a single sentence:[19] "7. ἐφόδιον, a Method, noticed by Suidas, who says that Theodosius wrote a commentary on it, but gives no further information about it." Suidas was a Greek encyclopedist who lived in the 10th century, when Theodosius (*c*. 160-90 B.C.) was a mathematician in Anatolia. But in 1899 Heiberg read about a palimpsest of mathematical content found in Constantinople. The word "palimpsest" means "scraped again." Normally it is a parchment that

[19] Heath, ed., *The Works of Archimedes*, p. xxxvii.

has been used two or three times, after being scraped or washed each time, due to a shortage of parchment or to its high price. This specific parchment contained a Euchologion written in the 12th, 13th, or 14th, century, over a mathematical manuscript of the 10th century. From a few specimen lines to which he had access, Heiberg suspected that it contained an Archimedian text. He traveled to Constantinople and examined the manuscript twice, in 1906 and 1908. Fortunately the original text had not been completely washed out and Heiberg was able to decipher much of the contents by inspecting the manuscript and taking photographs. The manuscript contained 185 leaves with Archimedes's works in Greek. Beyond the texts already known, it contained three treasures: (I) fragments of the *Stomachion*, (II) a large part of the Greek text of the work *On Floating Bodies*. (Until then it was believed to have survived only in the Latin translation made by Willem von Moerbeke in 1269 from a Greek manuscript which is now believed lost.) (III) Most of *The Method* by Archimedes! A work that had been lost for two thousand years (the last person to study it seems to have been Theodosius), of which we did not know even the contents, appeared out of nowhere, greatly expanding our knowledge about Archimedes. Even the comments on this work by Theodosius are no longer extant.

In 1907 Heiberg published the Greek text of *The Method*, and a German translation with commentary by Zeuthen. In 1912 Heath published a complement to his English translation of Archimedes's works, now including *The Method*. Between 1910 and 1915 Heiberg published a second edition of the complete works of Archimedes, in Greek and Latin, in three volumes. This second edition is much better than the first, and was republished in 1972.[20] Heiberg's discovery was featured on the first page of The New York Times in 1907.

But the story does not end here. In the period between 1908 and 1930 the manuscript disappeared, probably having been stolen. Around 1930 a French antiquities collector bought the manuscript, without the knowledge of the external world. In 1991 the collector's family put this manuscript on sale in an auction. Only then was it realized this was the manuscript discovered by Heiberg in 1906 and which was supposed to have been lost. In 1998 it was sold by Christie's, in New York. It was bought for 2 million dollars by an anonymous billionaire and lent to Walters Arts Gallery, of Baltimore, USA. A group of scholars, directed by Nigel Wilson and Reviel Netz, of Stanford University, are working on the restoration, digitization and publication of the manuscript, which

[20] Heiberg, *Archimedis Opera*

contains the only still surviving copy of *The Method*—a work that had been lost for 2,000 years!

The work's great importance is due to the fact that it contains practically the only report of a mathematician of antiquity describing the method he utilized in discovering his theorems. In all other surviving works we have only the theorems presented in final form, derived with a rigorous logic and with scientifically precise proofs, beginning with axioms and other theorems. This dry presentation conceals the method or the intuition that led to the final result. *The Method* changed all this. For here, Archimedes describes the path he followed to arrive at several significant results on quadrature and cubature (calculation of areas and volumes by integration), as well as the center of gravity of several important two- and three-dimensional geometric figures. Here are Archimedes's own words:[21]

> Archimedes to Eratosthenes greeting.
>
> I sent you on a former occasion some of the theorems discovered by me, merely writing out the enunciations and inviting you to discover the proofs, which at the moment I did not give. The enunciations of the theorems which I sent were as follows.
>
> (...)
>
> The proofs then of these theorems I have written in this book and now send to you. Seeing moreover in you, as I say, an earnest student, a man of considerable eminence in philosophy, and an admirer [of mathematical inquiry], I thought fit to write out for you and explain in detail in the same book the peculiarity of a certain method, by which it will be possible for you to get a start to enable you to investigate some of the problems in mathematics by means of mechanics. This procedure is, I am persuaded, no less useful even for the proof of the theorems themselves; for certain things first became clear to me by a mechanical method, although they had to be demonstrated by geometry afterwards because their investigation by the said method did not furnish an actual demonstration. But it is of course easier, when we have previously acquired, by the method, some knowledge of the questions, to supply the proof than it is to find it without any previous knowledge. This is a reason why, in the case of the theorems the proof of which Eudoxus was the first to discover, namely that the cone is a third part of the cylinder, and the pyramid of the prism, having the same base and equal height, we should give no small share of the credit to Democritus who was the first to make the assertion with regard to the said figure though he did not prove it. I am myself in the position of having first made the discovery of the theorem now to be published [by

[21] Heath, ed., *The Works of Archimedes*, Supplement, pp. 12-14.

the method indicated], and I deem it necessary to expound the method partly because I have already spoken of it and I do not want to be thought to have uttered vain words, but equally because I am persuaded that it will be of no little service to mathematics; for I apprehend that some, either of my contemporaries or of my successors, will, by means of the method when once established, be able to discover other theorems in addition, which have not yet occurred to me.

First then I will set out the very first theorem which became known to me by means of mechanics, namely that

Any segment of a section of a right-angled cone (*i.e.*, a parabola) is four-thirds of the triangle which has the same base and equal height,

and after this I will give each of the other theorems investigated by the same method. Then, at the end of the book, I will give the geometrical [proofs of the propositions]...

[I premise the following propositions which I shall use in the course of the work.] (...)

After this introduction about the life and work of Archimedes, we present several experiments that lead to a precise conceptual definition of the center of gravity of bodies.

Chapter 3

The Center of Gravity

3.1 Geometry

We begin our work with a little mathematics. We will cut out some plane figures and find their main geometrical properties. Later we will utilize these figures in experiments. The dimensions we present here are adequate for individual activities. Larger sizes should be used for demonstrations in the classroom, talks, and seminars.

3.1.1 Materials

Paper board, light cardboard, thick card, or pasteboard. Light wood/plastic/metal/styrofoam plane and rigid sheets can also be utilized.

White sheets of paper.

Ruler, pen, T-square and protractor.

3.1.2 Finding the centers of circles, rectangles and parallelograms

From a pasteboard we draw and cut out a circle 7 or 8 cm in diameter. If the circle is drawn with compasses, the center should be marked with a pen, and marked with an "X".

If the circle is drawn with a glass turned upside down, the center can be found by the intersection of two diameters. The diameters can be drawn with a ruler. But it is difficult to be sure if the ruler passes exactly through the center when we do not know exactly where the center is located.

An alternative procedure to find the diameter and center of the circle involves the paper. Later we will perform experiments with the pasteboards, so it is better not to fold them. For this reason the folding we discuss here should be done with similar figures made from sheets of

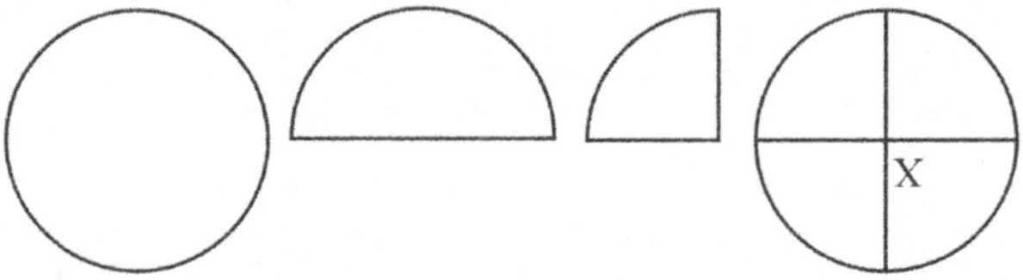

Fig. 3.1. Finding the center of a circle by paper folding.

paper. For example, we place the pasteboard circle on a sheet of paper and cut out a similar circle of paper. We then fold the paper circle in two equal halves. We fold it once more so that it is divided into four equal parts, as in Fig. 3.1. We can then use a pen to draw the diameters in the paper circle. The center of the circle is the intersection of the diameters. A hole should be made at the center. By placing the paper circle on the pasteboard circle, we can mark the center of the circle on the pasteboard.

We cut out a pasteboard in the shape of a rectangle with sides of 6 cm and 12 cm. There are two ways to find the center. The simplest one is to connect the opposite vertices. The center of the rectangle is the intersection of these diagonals, marked with the "X".

The other way is to find (with a ruler or by folding) the central point of each side. We then connect the middle points of opposite sides. The center is the intersection of these straight lines.

The parallelogram is a plane quadrilateral in which the opposite sides are parallel to one another. A parallelogram is cut out from a pasteboard with sides of 6 cm and 12 cm, with the smallest internal angle being 30° (or 45°). The center of this parallelogram can be found by the two methods we used for the rectangle, as in Fig. 3.2.

3.1.3 The triangle centers

There are three types of triangle: equilateral (three equal sides), isosceles (only two sides of the same length), and scalene (with three different sides). Every triangle has four special centers: circumcenter (C), barycenter or triangle centroid (B), orthocenter (O), and incenter (I). We will find these four special points in the case of an isosceles triangle with a base of 6 cm and height of 12 cm. With these dimensions each one of the equal sides has a length of 12.37 cm.

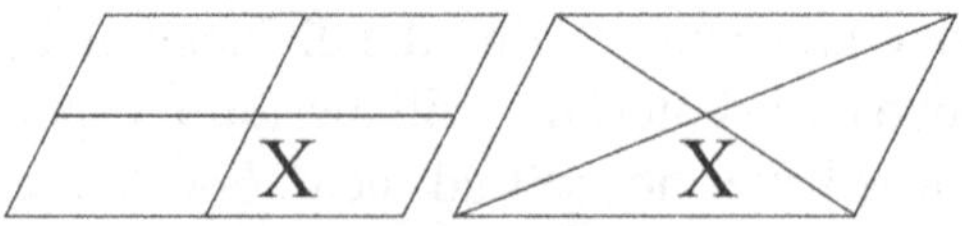

Fig. 3.2. Finding the center of a parallelogram.

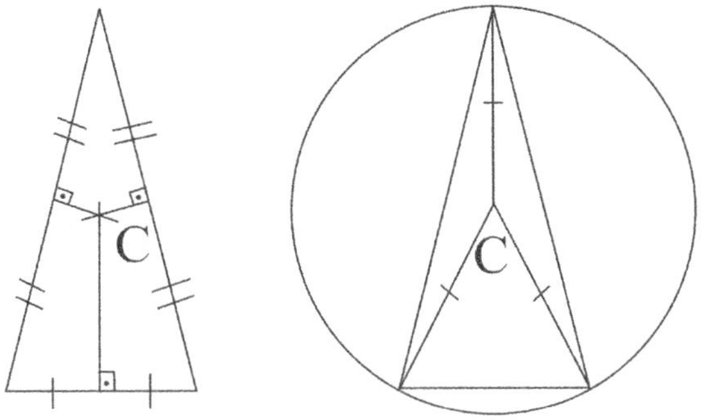

Fig. 3.3. The circumcenter and the circumcircle.

We draw and cut out a triangle of this size from a pasteboard. We also cut out four equal triangles from a sheet of paper. Each one of these four paper triangles will be used to draw the straight lines and locate one of the special points. When necessary, also the folding should be done with these paper triangles.

The circumcenter is the intersection of the perpendicular bisectors. A perpendicular bisector of a straight line AB is a straight line perpendicular to AB and passing through its midpoint M. To find the midpoint of each side we can use a ruler. With a T-square or using the pasteboard rectangle we draw a straight line perpendicular to each side through its midpoint. The intersection of these lines is the circumcenter (C), as in Fig. 3.3.

Another way of finding the midpoint of each side is by folding. In this case we only need to join the vertices two by two. The folding will be orthogonal to the side, passing through the midpoint.

An important property of the circumcenter is that it is equidistant from the vertices. It is therefore the center of the triangle's circumcircle, as in Fig. 3.3.

In every acute triangle (a triangle in which all angles are acute, that is, smaller than 90°), the circumcenter is inside the triangle. In a right-angled triangle the circumcenter is located at the midpoint of the hypotenuse. In every obtuse triangle (a triangle which has an obtuse angle, that is, larger than 90°), the circumcenter is outside the triangle.

The barycenter or triangle centroid (B) is the intersection of the medians, which are the lines connecting the vertices to the midpoints of the opposite sides. It is also called the median center. The midpoint of each side can be found with a ruler or by folding. After finding them, all you need to do is join these midpoints to the opposite vertices. The intersection of these medians is the centroid (B), as in Fig. 3.4. The bary-

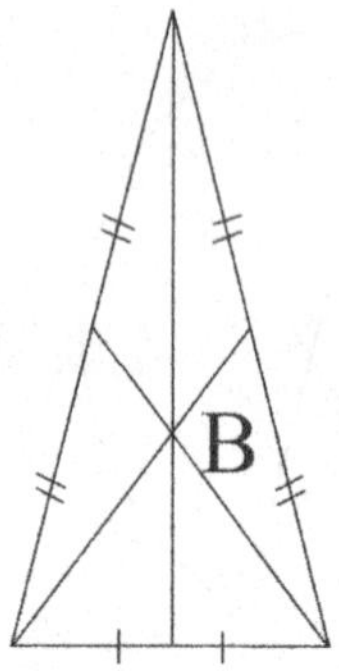

Fig. 3.4. The barycenter.

center is always inside the triangle and has an important property. The distance from the vertex to the centroid is always twice the distance from the centroid to the midpoint of the opposite side.

The orthocenter is the intersection of the altitudes of a triangle, which are the straight lines connecting the vertices to the opposite sides, orthogonal to them. The easiest way to find these lines is to use a T-square or pasteboard rectangle. We slide the base of the T-square or the rectangle along one leg of the triangle until the perpendicular side of the T-square or the rectangle meets the opposite vertex of the triangle. At this point we draw the perpendicular to the leg, connecting it to the opposite vertex, as in Fig. 3.5.

The orthocenter is the intersection of the altitudes, as in Fig. 3.5. The altitudes also represent the smallest distances between the vertices and the opposite sides. Depending upon the dimensions of the triangle, the orthocenter may be inside or outside the triangle.

The incenter is the intersection of the angle bisectors of the triangle, which are the straight lines dividing the vertices into two equal angles. These lines can be obtained with a protractor. But the easiest way

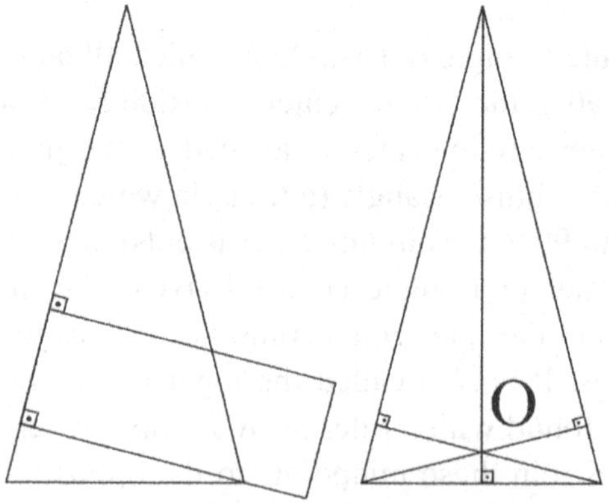

Fig. 3.5. The orthocenter.

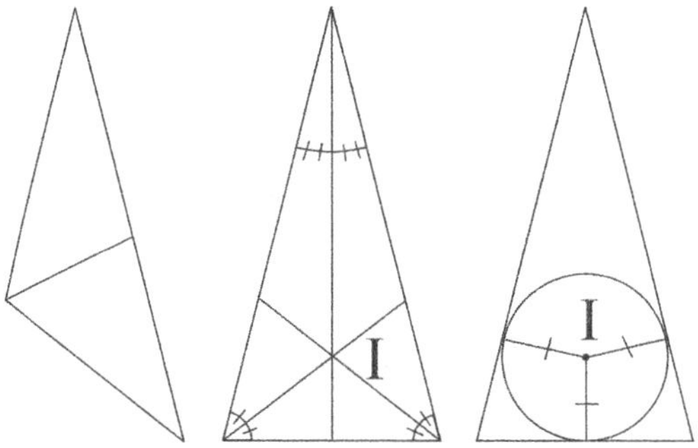

Fig. 3.6. The incenter and the incircle.

is by folding. In this case you only need to join the adjacent sides through the vertex. Folding divides each vertex into two equal angles. The intersection of the straight lines is the incenter (I), as in Fig. 3.6.

The incenter is always located inside the triangle. It is equidistant from all sides of the triangle. It is thus also the center of the incircle (the inscribed circle of the triangle, tangent to all three sides), as in Fig. 3.6.

After locating these four centers with the paper triangles, we make holes in the papers at these centers. We then superimpose each of these paper triangles upon the pasteboard triangle and mark these points. The final result in the case of an isosceles triangle with a 6 cm base and 12 cm height is shown in Fig. 3.7. We can see that these four points are different from one another, with the orthocenter closer to the base, then the incenter, the barycentre, and the circumcenter. These four points are along a straight line which is the angle bisector, altitude, median, and perpendicular bisector.

For an equilateral triangle these four centers superimpose on one another, as in Fig. 3.8.

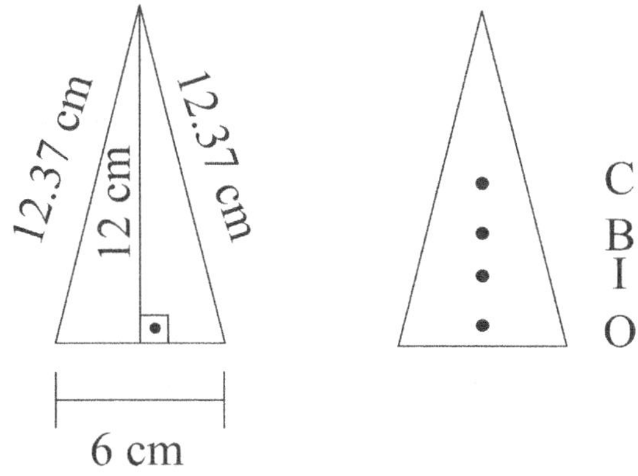

Fig. 3.7. Isosceles triangle and its centers.

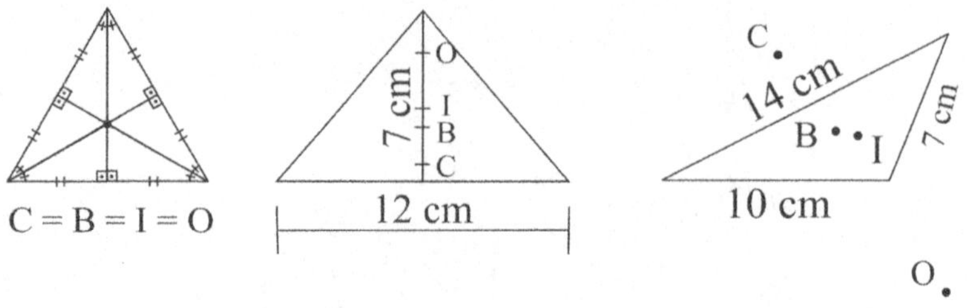

Fig. 3.8. The triangle centers in some special cases.

For an isosceles triangle with 12 cm base and 7 cm height the order of the centers relative to the base is opposite to the order for a 6 cm base and 12 cm height isosceles triangle, as in Fig. 3.8.

For a scalene triangle these four centers are not along a single straight line. Moreover, they are not all necessarily inside the triangle. In Fig. 3.8 we show an obtuse triangle with sides of 7 cm, 10 cm and 14 cm. We can see that the barycenter and the incenter are inside the triangle, while the circumcenter and orthocenter are outside it.

3.2 Experiments on and definition of the center of gravity

Thus far we have dealt only with geometry. Now we will begin to perform experiments. The majority of the experiments described here were inspired by the excellent works of Ferreira and Gaspar, highly recommended.[1].

We will use a few primitive concepts, that is, concepts that we cannot define without avoiding vicious circles. These are: body, relative orientation of bodies (body B located between bodies A and C, for instance), distance between bodies, change of position between bodies, and time between physical events.

Experiment

We hold a coin above the surface of the Earth and release it. We observe that it falls to the ground. The same happens with the pasteboard circles, rectangles and triangles.

This is one of the simplest and most important experiments of mechanics. Not all bodies fall to the ground when released in air. A helium filled balloon, for example, rises when released in air, moving away from the surface of the Earth. On the other hand, if it is released in a high vacuum it also falls to the ground. In this work we will perform

[1] N. Ferreira, *Mecânica*, Projeto RIPE – Rede de Instrumentação para o Ensino (São Paulo: Instituto de Física, USP); N. Ferreira, *Equilíbrio* (São Paulo: Instituto de Física, USP). Available in 2007 at: http://www.ludoteca.if.usp.br/; A. Gaspar, *Experiências de Ciências para o Ensino Fundamental* (São Paulo: Ática, 2003).

experiments in open air. All the bodies we analyze here fall to the ground when released at rest.

3.2.1 Definitions

We now define a few concepts that will be employed throughout this work.

Rigid body: Any body whose parts do not change their relative orientations and distances when this body moves relative to other bodies. The triangle pasteboard, for instance, can be considered as a rigid body for the purposes of this book. Even when the triangle falls and rotates relative the surface of the Earth, the parts of the triangle remain fixed relative to one another (the distance between any two points belonging to the triangle remain constant in time). On the other hand, a cat walking on the sidewalk cannot be considered a rigid body. The distance between its feet, or between a foot and the tip of the tail, does not remain constant in time. Most experiments in the first part of this book will be performed with rigid bodies. When we say "body," normally we refer to a "rigid body," unless specified otherwise.

Motion and rest: We say that two bodies A and B are in relative motion (or rest), when the distance between any particle i of body A and any particle j of body B does (does not) change with the passage of time. In this work we will often speak of the motion and rest of a body relative to the Earth. When we say simply that a body is at rest or in motion, we normally mean that it is at rest or in motion relative to the surface of the Earth. The same should be understood for all the parts of a body in relation to all the parts of the Earth.

Equilibrium: We will normally understand equilibrium of a body as its state of rest relative to the surface of the Earth. That is, when we say that a body is in equilibrium, we mean that all of its parts remain at rest relative to the Earth with the passage of time. When a triangle is in our hands, we say that it is in equilibrium. When it is falling to the ground, it is no longer in equilibrium.

Gravity: Name given to the property which makes the bodies fall toward the surface of the Earth when released at rest. This can also be expressed by saying that gravity is the tendency of bodies to be attracted toward the center of the earth.

Go down and up: When we say that a body is going down (up), we mean that it is moving toward (away from) the surface of the Earth. Instead of these verbs we can also employ analogous terms, like fall and rise.

On top and bottom, above and below: When we say that body A is above body B, we mean that B is between the Earth and body A. When we say that body A is below body B, we mean that A is between the Earth and body B. When we refer to the top (bottom) part of a body, we mean its part farthest (closest) to the surface of the Earth.

Vertical: Straight line defined by the direction followed by a small dense body (like a metal coin) when it falls toward the Earth due to the action of gravity, beginning from rest. It is also the straight line followed by a body which moves upward when released from rest (like a helium balloon, in a region without wind). That is, the vertical is not an arbitrary straight line. It is a very specific straight line connected with the Earth's gravity. Here we are neglecting the influence of wind.

Horizontal: Any straight line or plane orthogonal to the vertical line.

It should be stressed that all these concepts are connected to the Earth, indicating physical properties related to the gravitational interaction of the bodies with the Earth. That is, they are not abstract or purely mathematical concepts. They are defined beginning from mechanical experiments performed at the surface of the Earth.

It is important to introduce all these concepts explicitly because they will be utilized throughout this book. Nevertheless, it should be stressed that they are idealizations which are never found exactly like this in nature. For example, no body is perfectly rigid. Even when a book is resting above a table, its molecules are vibrating. In this sense, no body is actually in equilibrium according to the definition above, as parts of this body will always be moving relative to the surface of the Earth, even when the body as a whole, macroscopically, is at rest. When we support a body from below with a stick, the body will suffer a small curvature, even if it is a metal plate. However, for phenomena at a macroscopic scale, these details (the vibration of the molecules, or the small curvature of the body) are not easily observable, or may not be relevant for the case under consideration. For this reason the concepts already defined make sense at the macroscopic scale and should be understood as such.

3.2.2 Support for the experiments

After these definitions we can go on with the experiments. We concentrate on the phenomena leading to the definition of center of gravity. To this end we will need a stand to support the plane pasteboard figures we made earlier. There are several ways to make a stand.

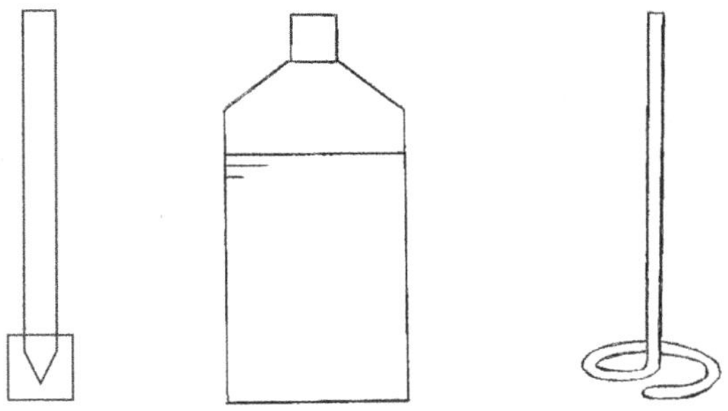

Fig. 3.9. Supports for the experiments.

A bamboo barbecue skewer: We use children's modeling clay as a base and fix the bamboo skewer vertically, with the tip down, as in Fig. 3.9. It is important to stress that the tip should be pointing toward the Earth; otherwise it will be very difficult to perform the equilibrium experiments. The bamboo skewer can also be fixed in rubber or another appropriate base.

A pencil stand: We fix a pencil vertically in a sharpener, with its tip down.

A bottle stand: If the pasteboard figures are large (with lengths of the order of 20 cm or of 40 cm, an appropriate size for demonstrations in the classroom), you can use a glass or plastic bottle as a stand, with the pasteboard resting above the cover. If the bottle is made of plastic, it should be water filled to prevent if from falling over during the experiments, as in Fig. 3.9.

A wire stand: Another interesting possibility is to utilize a thick, solid vertical wire with a spiral base, as in Fig. 3.9. If the wire is rigid but thin, it may be difficult to balance the figures horizontally above it. Moreover, the wire could pierce a hole in styrofoam sheets, *etc.* As a result, a thick, rigid wire is preferred.

A nail stand: In this case we only need a nail fixed in a cork, rubber, wooden board, or other convenient base. The head of the nail should be horizontal, with the point fixed in the base.

There are many other possibilities. The important points are that the stand should be rigid, fixed in an appropriate base, and remain vertical, and its top should be flat and remain in a horizontal plane. Moreover, the size of the top should be small compared with the dimensions of the figure which that is balanced on it. But it cannot be extremely small, or pointed (such as, for instance, the bamboo skewer, pencil or nail with the tip pointing upward). The top end has to be small in order

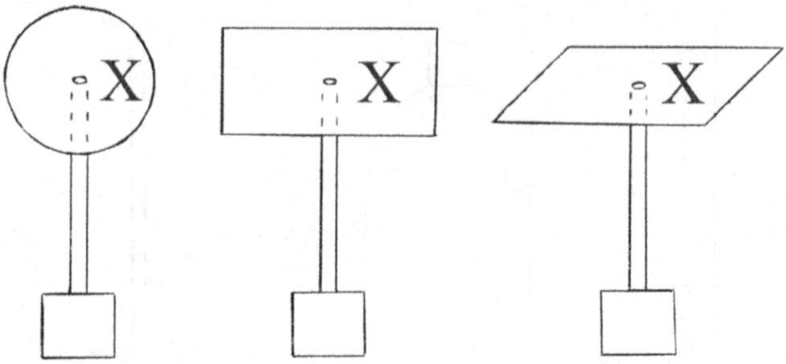

Fig. 3.10. The circle, rectangle and parallelogram only remain at rest when supported by their centers.

to clearly locate the equilibrium point of the body, but should not be too small, otherwise many of the experiments will be impracticable. (If it has a negligible area, it is very difficult to keep a body at rest above it.) With a little practice we can easily find appropriate dimensions.

3.2.3 First experimental procedure to find the CG

Experiment

We try to balance the circle, rectangle, and parallelogram pasteboards in a horizontal plane by supporting them on a vertical stand. We take the circle, for example, lay it horizontal, and place it with the stand under one of its points, releasing the circle from rest. We observe that it always falls to the ground except when the support is under the center of the circle. With all these plane figures we observe that there is a single point in each one which must be on the vertical stand in order for the figure to remain at rest after release. Experience teaches us that for the rectangle and parallelogram, this special point is also the center of these figures, as happened with the circle, as in Fig. 3.10.

As an historical curiosity, it is worth noting that Archimedes was the first to prove theoretically that the center of gravity of a circle is its center, and that the center of gravity of a parallelogram is the intersection of its diagonals (rectangles and squares are particular cases of parallelograms). Lemma 6 of *The Method* says: "The center of gravity of a circle is the point that is also the center [of the circle]."[2] Proposition 9 of his work *On the Equilibrium of Planes* states: "The centre of gravity of any parallelogram lies on the straight line joining the middle points of opposite sides."[3] And finally, proposition 10 of the same work states:

[2] Heath, ed., *The Works of Archimedes*, p. 15.
[3] Heath, ed., *The Works of Archimedes*, p. 194.

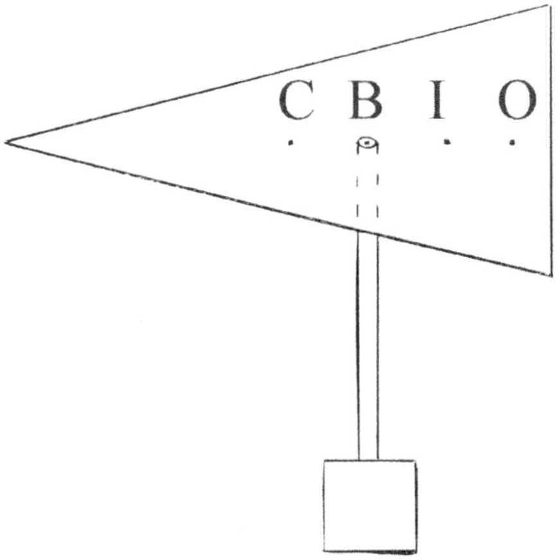

Fig. 3.11. We can only equilibrate an horizontal triangle by supporting it through its barycenter.

"The centre of gravity of a parallelogram is the point of intersection of its diagonals."[4]

These bodies are balanced only when the stand is under their centers. This equilibrium is connected with the Earth's gravity. Our first reaction would be to call the centers of these bodies their "centers of gravity." But from the result of the next experiment and its analysis, we will see that this definition has to be modified. But for the time being we can say from the experiments performed thus far that only when these specific bodies are supported by their centers do they remain in equilibrium when released from rest. We thus give a first provisional definition.

Provisional definition CG1: We call the center of gravity of a body its geometric center. This point will be represented by the letters CG.

Experiment

We now equilibrate an arbitrary triangle (equilateral, isosceles or scalene) in a horizontal plane above a vertical stand. As a concrete example we will consider the pasteboard isosceles triangle of base a and height b ($a = 6$ cm and $b = 12$ cm). This triangle has its four special centers (orthocenter, circumcenter, barycenter and incenter) well separated from one another. We utilize now a barbecue bamboo skewer as the vertical stand. In this way we can locate clearly the equilibrium point of the triangle. That is, the point below which the bamboo skewer should be placed in such a way that the triangle remains in equilibrium, after placed in a horizontal plane and released from rest. Experiment teaches that the triangle always falls to the ground, except when supported by the barycenter, as in Fig. 3.11. Even when supported by the circumcen-

[4] *Ibid.*, p. 195.

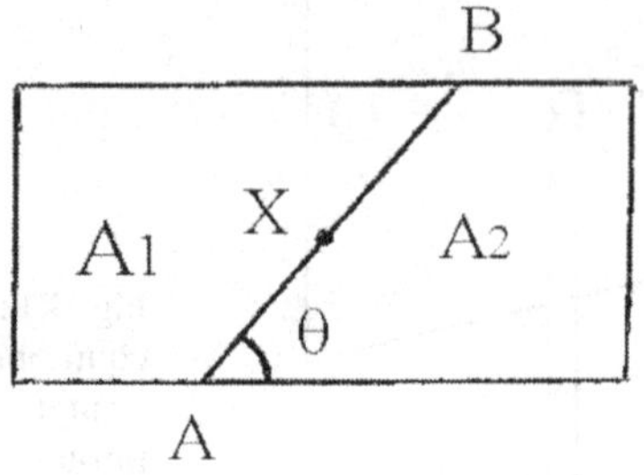

Fig. 3.12. The geometric center X of a rectangle: The segment $AX = XB$ and the area $A_1 = A_2$ for any angle θ.

ter, by the orthocenter, by the incenter, or by any other point (except the barycenter), the triangle always falls after release from rest.

Once more Archimedes was the first to prove theoretically that the center of gravity of any triangle coincides with the intersection of the medians. Proposition 13 of his work *On the Equilibrium of Planes* reads: "In any triangle the centre of gravity lies on the straight line joining any angle to the middle point of the opposite side."[5] Proposition 14 states: "It follows at once from the last proposition that the centre of gravity of any triangle is the intersection of the lines drawn from any two angles to the middle points of the opposite sides respectively."[6]

Can we say that the barycenter of a triangle is its geometric center? Does every triangle have a geometric center? In order to answer these questions we need to know what we mean by "geometric center." Intuitively we think of a geometric center as a point of symmetry of the body. In order to quantify this qualitative idea of symmetry, we can think of the center X of a rectangle. Let us consider a straight line AXB passing by X, making an angle θ with the base, and dividing the rectangle into two parts of areas A_1 and A_2, as in Fig. 3.12.

There are two criteria by which we can say that X is the geometric center of the rectangle. (I) The straight line AXB is always divided in two equal segments by X. That is, AX = XB for every angle θ. (II) The straight line AXB always divides the rectangle into two equal areas. That is, $A_1 = A_2$ for any angle θ. These two properties will not be valid for any other point of the rectangle—only for its center X. Let P be another point of the rectangle. A straight line APB may be divided in two equal segments AP = PB when it is inclined by a specific angle θ_1 relative to the base of the rectangle, but this will no longer be valid when we

[5] Heath, ed., *The Works of Archimedes*, p. 198.
[6] Ibid., p. 201.

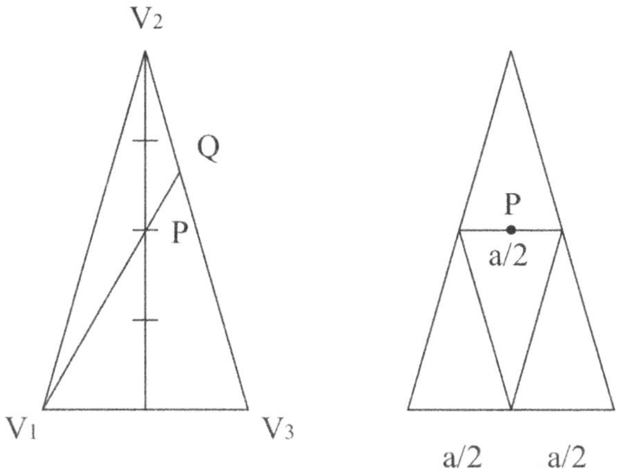

Fig. 3.13. Criteria (I) and (II) will not be true for any point P of a triangle.

change the angle θ_I. Another straight line CPD may divide the rectangle into two equal areas when it is inclined by an angle θ_{II} relative to the base of the rectangle. But once again, this will not be valid when we change the angle θ_{II}. We then conclude that the rectangle has only a single geometric center, the same being true of a circle and some other symmetric figures, such as a parallelogram or an ellipse.

On the other hand, criteria (I) and (II) in the previous paragraph will not be true for any point P of a given triangle. That is, given an arbitrary triangle, there is no point P_I belonging to it such that all straight lines passing through P_I will satisfy criterion (I). Moreover, there is no point P_{II} belonging to it such that all straight lines passing through P_{II} will satisfy criterion (II). In this sense we can say that no triangle has a geometric center. On the other hand, every triangle has four special centers (circumcenter, barycenter, orthocenter, and incenter).

In order to illustrate this fact we consider the isosceles triangle $V_1V_2V_3$ with base a and height b. The area of this triangle is $ab/2$. The median connecting the center of the base to the vertex V_2 is divided into two equal parts by a point P located at a distance $b/2$ from the base and from the vertex V_2. A straight line parallel to the base and passing through P and limited by the sides of the triangle is also divided into two equal parts by P. On the other hand, the straight line V_1PQ (where the point Q is the intersection of the side V_2V_3 with the extended line V_1P) is not divided into two equal parts by P, as in Fig. 3.13. That is, criterion (I) is not satisfied by P. The same is true for any other point on the triangle.

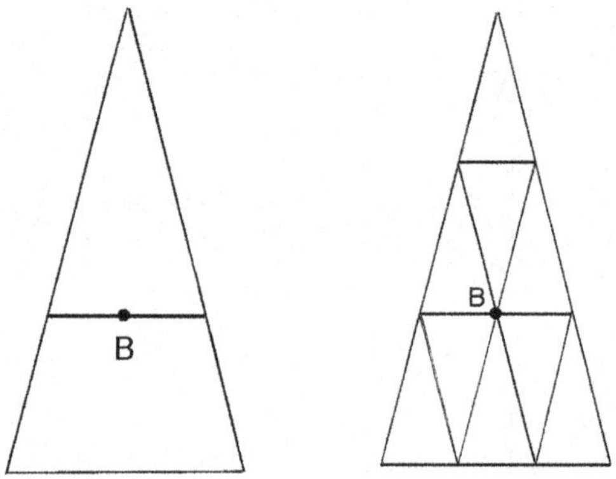

Fig. 3.14. A straight line parallel to the base and passing through the barycenter divides the triangle into two figures having different areas.

Nor is criterion (II) satisfied by P. The straight line passing through V_2 and P divides the triangle into two parts of equal area. On the other hand, the straight line parallel to the base and passing through P does not divide the triangle into two parts of equal areas, as in Fig. 3.13. The upper triangle has only a quarter of the total area, while the lower trapezium has three quarters of the total area.

The barycenter B is located at a distance $b/3$ from the midpoint of the base and at a distance $2b/3$ from the upper vertex. This shows at once that it does not satisfy the previous criterion (I). The extended straight lines connecting B to any one of the vertices divide the triangle in two parts of equal areas. But this will not be the case, for example, for a straight line parallel to the base and passing through B, as in Fig. 3.14.

In this case the area of the upper triangle is equal to $4/9$ of the total area, while the area of the lower trapezium is equal to $5/9$ of the total area. In order to confirm this without performing the calculations, all we need to do is cut out nine equal isosceles triangles, each with a base of $a/3$ and height of $b/3$ (area of ab/18). We can fill the superior triangle with four of these small triangles, and the inferior trapezium with five of these small triangles, as in Fig. 3.14.

Even the most symmetrical triangle, the equilateral triangle, has no geometric center that satisfies criterion (I) or criterion (II). In this case the four special centers coincide at the barycenter B of the triangle. And we just saw that the barycenter of an isosceles triangle does not satisfy any of these criteria. As the equilateral triangle is a particular case of an isosceles triangle, it follows automatically that the barycenter of an equilateral triangle will not satisfy any of these criteria. Nevertheless, we can say that the equilateral triangle has a center of symmetry given by C = B = O = I. Although this point does not satisfy criteria (I) and (II), there is symmetry of rotation around an axis orthogonal to the plane of the triangle and passing through this point. That is, any characteristic of the triangle is repeated with a rotation of 120° around this point. For this reason it is possible to say that the barycenter of an equilateral triangle is its center of symmetry, yet not its geometric center.

We then conclude that a triangle has no geometric center defined according to criteria (I) and (II). Nevertheless, experience teaches us that every triangle can be balanced horizontally when supported by a thin vertical stand placed under its barycenter, but not when we place the support under any other point of the horizontal triangle. This suggests that we should change our previous definition of center of gravity. We now give a second, more precise provisional definition of a CG.

Provisional definition CG2: The center of gravity is a certain point in the body such that if the body is supported by this point and released from rest, it remains in equilibrium relative to the Earth.

Later on we will need to change this definition yet again for a more general concept. But for the time being it is a suitable definition. From the experiments performed thus far it follows that any body has only a single point satisfying this definition. If the body is released from rest when supported by any other point, it does not remain in equilibrium, but falls to the ground. For circles and parallelograms the CG is the center of these bodies, while for the triangles it coincides with the barycenter.

Another way of looking at CG has to do with the weight of the bodies. Later on in this book we will quantify this concept and show how it can be measured. But we all have an intuitive notion of the weight of a body as a quantitative measure of gravitational force. We say that body A is heavier than body B when it is more difficult to keep A at a certain height from the ground than it is to keep B at rest at the same height. This difficulty can be indicated by our sweat, by the fatigue we feel in an outstretched arm, or by the deformation created by bodies A and B upon the body supporting them (in the case of a flexible support like a spring, for example).

In the previous Figures we saw that the whole weight of the circle, rectangle, parallelogram or triangle was supported by the bamboo skewer placed at a single point below each of these bodies. We can then give a new provisional definition of the CG.

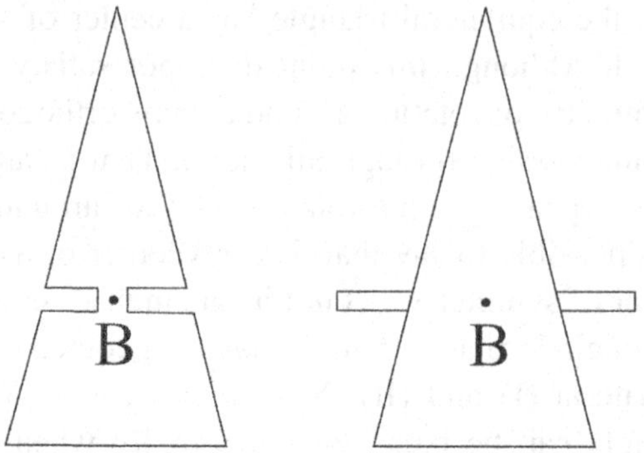

Fig. 3.15. Although the upper triangle and the lower trapezium have different areas and weights, this figure can be kept in horizontal equilibrium by a support passing through the barycenter.

Provisional definition CG3: The center of gravity of a body is the point of application of the gravitational force acting upon it. That is, it is the point of the body upon which all the gravity acts when it is at rest, the point where the weight of the body is located or concentrated. It can also be called the center of weight of a body.

As we have seen, a triangle has no geometric center. This leads to an important conclusion which will be explored in the following experiment.

Experiment

We have seen that not all straight lines passing through the barycenter of a triangle divide it into two equal areas. As we are dealing with homogeneous plane figures, the weight of any part is proportional to its area. This fact suggests a very interesting experiment. We cut out a pasteboard triangle with base a and height b (for example, with $a = 6$ cm and $b = 12$ cm). The barycenter is located along the median connecting the superior vertex to the midpoint of the base, at a distance $2b/3$ from the superior vertex. We can then cut this triangle in two parts with a pair of scissors, cutting a straight line parallel to the base and passing through the barycenter. We then connect the two parts only by the central region around the old barycenter utilizing a small piece of pasteboard. Alternatively, we can remove two narrow strips parallel to the base on either side of the barycenter, keeping only a small region around the barycenter, as in Fig. 3.15.

We then try to equilibrate this figure horizontally above a vertical stand. We observe that the body only remains balanced horizontally when the stand is placed below the barycenter. That is, although the area and weight of the trapezium are larger than the area and weight of the small triangle (which goes from the superior vertex to the straight line passing through the barycenter), the system as a whole remains in equilibrium. If these two parts were not rigidly connected, each one of them would fall to the ground after release. We then conclude that the CG is not, necessarily, the point which divides the body in two equal areas or in two equal weights. We will discuss this aspect in more detail in a later section.

Experiment

There is another way to perform this experiment without cutting the larger triangle. We take the original triangle of base a and height *b* and balance it horizontally by placing the triangle on the edge of a ruler in the vertical plane. The edge of the ruler should be parallel to the base of the triangle, passing through its barycenter. The extended vertical plane passing through the ruler divides the triangle into two different areas, that is, into two different weights. Nevertheless, the triangle remains in equilibrium when supported by this ruler, as in Fig. 3.15.

3.2.4 Experiments with concave bodies or pierced bodies

We now cut out some concave figures, such as a letter C, a first quarter Moon, a boomerang, *etc.* Some pierced bodies should also be prepared, such as a washer (a metal washer is easily found). To cut the interior circumference of a pasteboard washer, a radial cut can be made between the exterior and interior circles. But with a pair of pointed scissors this is unnecessary. The outside diameters of these figures can be 8 cm or 10 cm, for instance. The interior diameters can be of the order of 4 cm or 6 cm. But these dimensions are not so important. For the following experiments you will need to cut out at least two equal figures of each model (two letters C of the same size and shape, two Moons, two washers, *etc.*). One set of these figures will be used in the next experiment, while the other set of identical figures will be used in later experiments (with sewing threads ttached to these figures with adhesive tape).

Experiment

We try to balance these figures in a horizontal plane by placing them above a vertical stand, as we did with the triangle. We observe that we cannot balance any of them. They always fall to the ground, no matter where we place the support, as in Fig. 3.16.

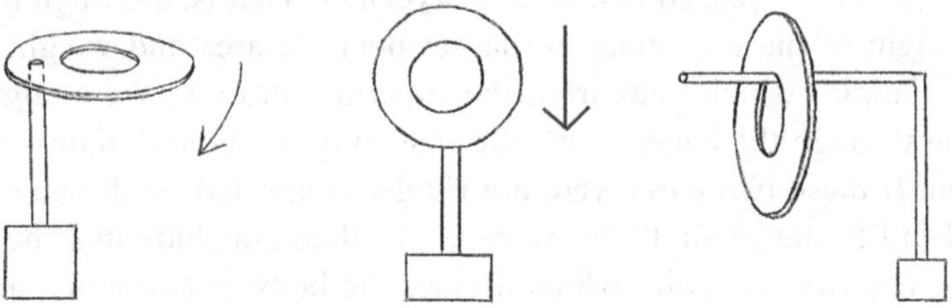

Fig. 3.16. The washer falls to the ground when we try to support it horizontally or vertically by an edge. But we can support it by a horizontal bamboo skewer passing through its hole.

Even when we try to balance them on an edge, by placing the figures in a vertical plane above the support, we do not succeed; they always fall to the ground, as in Fig. 3.16.

The only way to balance them is to hold the bamboo skewer horizontally and the figures in a vertical plane, with the bamboo skewer passing through a hole in the bodies, or supporting the concave part of the figures, as in Fig. 3.16.

There are different ways to analyze this experiment. The first is to conclude that some concave or pierced bodies do not have a specific center of gravity, but do have an entire line of gravity. The washer, for instance, can remain balanced in a vertical plane when supported by any point belonging to its interior circumference. On the other hand it cannot be balanced when the vertical bamboo skewer is placed exactly at the empty center of the washer, which is its geometric center. If we follow definition CG2 rigorously, we must say that the washer has a line of gravity, namely, its interior circumference. In this case we cannot say that it has a specific CG located at a single point.

The same can be said of definition CG3. After all, the horizontal bamboo skewer in Fig. 3.16 is holding the whole weight of the vertical washer supported at a point on the interior circumference of the washer. But a vertical bamboo skewer cannot support a horizontal washer when the end of the bamboo skewer is located at the empty center of the washer. If we follow definition CG3 rigorously, we should say that the washer has a line of weight or a line of gravity, but not a specific point that could be called its CG.

Another way to analyze this experiment is to say that the CG does not need to be located "in the body." That is, we may say that it does not need to be located at any point coinciding with a material part of the body. The CG might then be located in empty space, at a point in some definite spatial relation with the body (like the geometric center of the washer, for instance), even though not physically connected to the body.

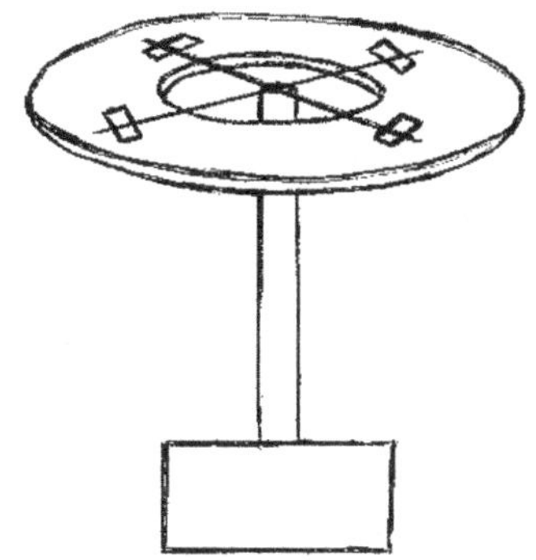

Fig. 3.17. The washer can be supported by its center utilizing two stretched sewing threads.

If we adopt this alternative, we will need to change our definition CG2. We will also need to find another way of experimentally locating the CG in these special cases, as presented in the procedure of the next experiment.

Experiment

We attach two taught sewing threads to the washer with adhesive tape, as if they were two diameters intersecting each other at the center of the washer. Now we can balance the washer in a horizontal plane by placing the vertical support under the intersection of the threads, as in Fig. 3.17. We can also find a similar point for the Moon or for letter C, by trial and error, *i.e.*, the intersection of two taught threads attached to the figures so that they remain in equilibrium horizontally when the support is placed vertically under this intersection.

If we accept this second alternative, we need to generalize our definition CG2 to include these special cases. A more general definition is presented now.

Provisional definition CG4: We call the center of gravity of a body a point in the body or outside it such that, if the body is supported by this point and released from rest, it remains in equilibrium relative to the Earth. When this point is located outside the body, a rigid connection must be made between this point and the body in order for the body to remain in equilibrium when released from rest.

This definition has a problem. After all, when we make this material rigid connection (like the taught threads attached with adhesive tape) the original body has been modified. But provided the weight of this material connection is small in comparison with the weight of the body, this is a reasonable procedure.

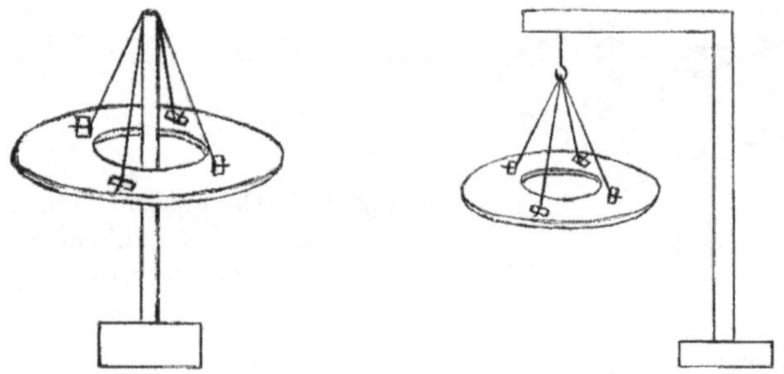

Fig. 3.18. The washer can also be supported along its axis of symmetry utilizing loose threads.

Even so there is another problem with this definition, as we will see in the next experiments.

Experiment

We now attach two loose threads of the same length to the washer with adhesive tape. The length of each thread should be greater than the external diameter of the washer. They are attached as in the previous experiment, at the same locations, *i.e.*, the straight line joining the two pieces of adhesive tape attaching each thread passes through the center of the washer. The only difference is the length of the threads. In this case we can also balance the whole system on a support. But now the intersection of the two threads touching the support is along the axis of symmetry of the washer, as in Fig. 3.18. It is no longer located at its geometric center.

If we follow the second alternative discussed above (where the CG does not need to be located in the body, and could be located in empty space), we must conclude that the washer has not just one center of gravity, but an infinite set of them located along its axis of symmetry. That is, the whole axis of symmetry of the washer might be called its "axis or line of gravity." This would be true both according to definition CG3, and according to definition CG4.

Definition CG3 has also problems with concave or pierced bodies. According to this definition the CG is the point of application of the gravitational force, that is, the point where gravity acts. The problem with this definition is that we normally consider gravity to be a force acting upon material bodies due to an interaction between the body and the Earth. It would be difficult to say that the point of application of the gravitational force on a washer was acting on the empty space where its geometric center is located. The force of the Earth cannot act on empty space. As a result, definition CG3 will have to be modified. It might, for instance, go something like the following.

Provisional definition CG5: The center of gravity is one point inside or outside the body which behaves as if all gravitational force where acting at this point. When this point is located outside the body, there must be a rigid connection between this point and the body in order to perceive or measure the entire gravitational force acting at this point.

This is a very reasonable definition. The greatest difficulty we encounter when dealing with it is how to locate this point. Let us consider the washer with loose threads, for instance. It is supported by four pieces of adhesive tape attached to it. These pieces of tape are supported by two taught threads. These threads, meanwhile, are supported at their intersection by a base or hook. That is, the washer behaves as if all its weight were supported along its axis of symmetry, at the intersection of the threads, away from the geometric center of the washer, provided we utilize threads attached to the washer. So it would make more sense to talk of a line of gravity, or a line of weight, instead of a center of gravity or a center of weight.

In the next experiments we will see another problem that arises even with the more general definitions CG4 and CG5.

3.2.5 Experiments with three-dimensional bodies

Thus far we have performed experiments with "plane" figures, or two-dimensional bodies. However, every material body is three-dimensional. When we say that a figure is plane or two-dimensional, we mean that its thickness is much smaller than the other dimensions involved in the problem (the thickness d of the pasteboard rectangle, for example, is much smaller than the length of its sides a and b). We now perform experiments with bodies in which all three dimensions are of the same order of magnitude.

The bodies we will consider are a cube or die with plane faces, a sphere, a metal screw-nut and an egg. For lighter bodies we use children's modeling clay and the barbecue bamboo skewer as support. For the egg (and other heavy spheres) we can use the table as a support, since it only touches the table in a small region due to its convex shape at all points.

Experiment

We release these bodies upon a horizontal support and observe the points at which they remain in equilibrium. In the case of the cube we

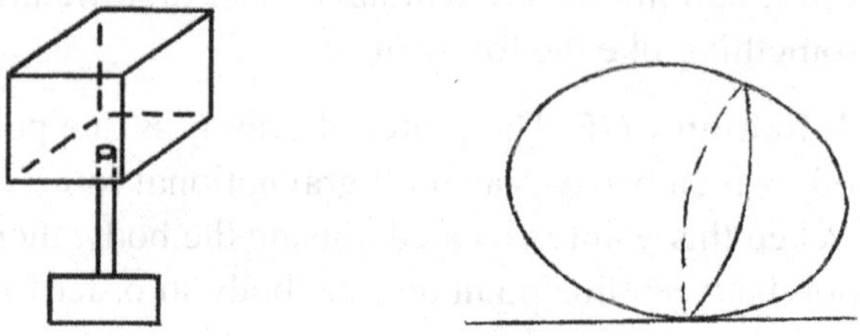

Fig. 3.19. A cube and an egg.

find six points of equilibrium, namely, the centers of the faces, as in Fig. 3.19.

For the metallic screw-nut we also find six points of equilibrium, the centers of the six external faces. Moreover, using the procedure involving the intersection of sewing threads (which we used above with the washer), it can be shown that all points along the axis of symmetry are also points of equilibrium of the screw-nut. It also remains balanced by any point along the internal circumference or cylinder surface if the barbecue bamboo skewer is fixed in a horizontal position, as was the case with the washer.

The sphere remains in equilibrium at all points of its surface. Therefore it has an infinite number of equilibrium points.

The most interesting case is that of the egg, which has a whole line of equilibrium. This line forms a circumference on the shell, such that the plane of this circumference is orthogonal to the axis of symmetry of the egg, as in Fig. 3.19.

From this experiment we conclude that many geometric bodies have more than one center of gravity if we follow definitions CG2, CG3, CG4 or CG5. The cube, for instance, would have six centers of gravity, the egg would have a whole line and the sphere its entire surface. The screw-nut would have six of these centers, the centers of the external faces, in addition to its internal circumference and to all points along its axis of symmetry. In order to be consistent with this discovery we should talk of points, lines or surfaces of gravity, instead of speaking of a single "center" of gravity for each body.

Fortunately there is another experimental procedure involving gravity with which we can find a single and specific point in each rigid body related to its condition of equilibrium relative to the Earth. By utilizing this second experimental procedure we can obtain another definition of the CG which avoids the previous problems and which has a relevant physical meaning. As this new procedure employs a plumb line, we first explain the instrument.

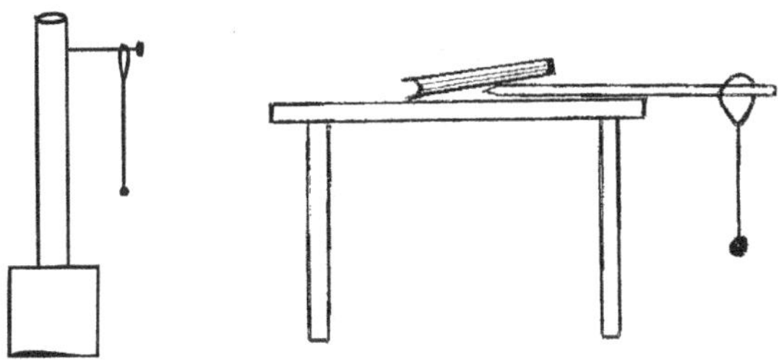

Fig. 3.20. Plumb line.

3.2.6 Plumb line, vertical and horizontal

Plumb line: This is the name given to any thread fixed at its upper end (this end remains at rest relative to the Earth) and which has a body fixed at its lower end. The plumb line must be free to oscillate around the extreme upper end.

The point of suspension, represented in some Figures by the letters PS is the point at which the body is hanging or suspended, as we will see in the next experiments. (Often it will coincide with the location of the pin or needle holding the body and the plumb line.)

The stand or aid point, represented in some Figures by the letters PA, is the upper end of a stand on which the body is supported, like the end of the bamboo skewer used as a support in some of the previous experiments.

The upper end of the plumb line can be held by our fingers, or tied to a bar or a hook, *etc.* In our experiments we will fix this top end to a rigid support at rest relative to the Earth. We stick a pin or needle into the upper part of our bamboo skewer placed in a vertical position, as in the experiments performed earlier. On the pin we will hang pierced pasteboard figures and also a plumb line. The plumb line will be a sewing thread with a weight at the bottom. We could simply tie or fasten it to the pin, but we will need to remove and replace the plumb line on the pin several times. Therefore it is best to make a small loop at the top of the thread. At the bottom of the thread we tie a plumb or a piece of modeling clay. The device to be used in the experiments is shown in Fig. 3.20.

One of the advantages of this device is that it allows us to repeat the previous experiments in which we supported pasteboard figures horizontally on a vertical bamboo skewer. In order to avoid the hindrance of the pin (touching the figure placed horizontally above it), the pin should be stuck in a little below the end of the bamboo skewer. In

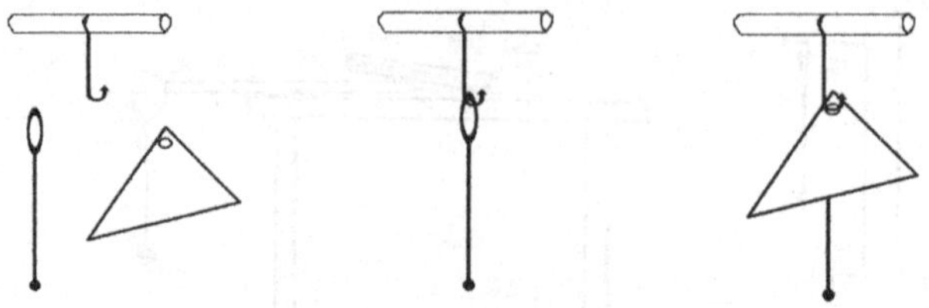

Fig. 3.21. Support with hook, plumb line and pierced pasteboard figure.

addition, the pin should not be perfectly horizontal, but inclined with its head a little above the point stuck in the bamboo skewer, in order to prevent the pasteboard figures from sliding off.

If we wish to perform experiments only with the plumb line, a bamboo skewer should be tied to it, in a horizontal position. This way we avoid the pins, which can be dangerous if we perform these experiments with children. The bamboo skewer is laid on a table, half of it extending beyond the table. The part on the table is kept in its place by a book or other weight on it. The plumb line hangs from the part of the bamboo skewer outside the table, free to oscillate, as shown in Fig. 3.20. The pierced pasteboard figure will also hang from the bamboo skewer, instead of being suspended by the pin.

Another practical alternative is to use a thread or lace tied to a bar or broomstick fixed in a horizontal position. At the bottom of the thread we attach a hook, from which we will hang the plumb line and the pierced pasteboard figures, as in Fig. 3.21.

Experiment

We hang the plumb line from the support and wait until it reaches equilibrium. Then we release a coin from rest close to the plumb line. We observe that the direction of fall is parallel to the plumb line, as in Fig. 3.22.

This is the main function of a plumb line. When it is at rest relative to the Earth, it indicates the vertical direction. In this sense it is a better indicator than a falling body, as it has a visible line, permanent and stable (when there is no wind blowing, *etc.*) Bricklayers often use plumb lines to determine whether a wall is vertically true.

There are three principal methods for finding the horizontal direction.

A) We first find the vertical, V, with a plumb line. Then we place a large T-square parallel to the plumb line. The direc-

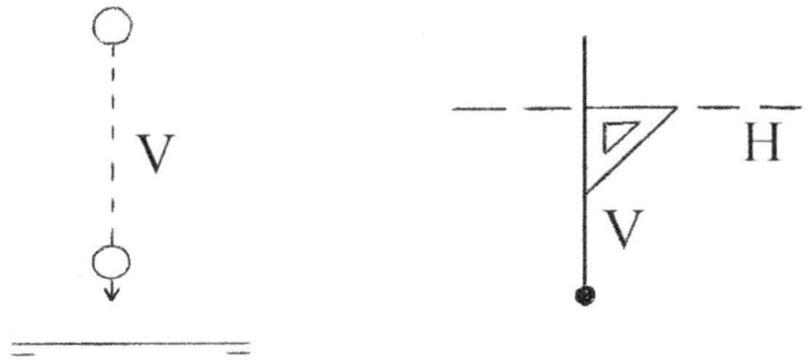

Fig. 3.22. Finding the vertical (V) and the horizontal (H) with a falling body or with a plumb line.

tion orthogonal to the line indicated by the T-square is then, by definition, the horizontal direction, H, as in Fig. 3.22.

B) With a spirit level. Usually it has the shape of a small parallelepiped with an internal cylindrical transparent vessel containing a liquid with a bubble. There are two straight marking lines along the axis of the cylinder, symmetrically located relative to the center. The spirit level is placed on a surface. When the bubble remains in the middle of the two marks the surface is horizontal, as in Fig. 3.23.
When the bubble remains at one of the ends of the vessel, the surface is not horizontal. The side where the bubble is located is higher than the opposite side. The spirit level works due to the action of gravity and the upward thrust exerted in a fluid (the principle of Archimedes).

C) We use a transparent hose open at both ends and partially filled with a liquid, such as water. The hose is kept at rest relative to the Earth and we wait until the liquid reaches equilibrium. The straight line connecting the two free surfaces of liquid indicates the horizontal direction, as in Fig. 3.23. It works based upon the equilibrium of liquids under the action of gravity.

As a curiosity it is worth mentioning here how bricklayers build orthogonal walls. After finishing a wall, they mark two points on it 4 m

Fig. 3.23. Finding the horizontal with a spirit level and with a transparent hose.

apart horizontally, A and B. The first point, A, is at the end of the wall where the other wall is to be built. Next they find a third point C such that the distance between A and C is 3 m and the distance between B and C is 5 m. The straight line connecting AC is then orthogonal to the straight line connecting AB. Instead of these specific distances, any multiple of them can be used (30 cm, 40 cm and 50 cm, for instance). The principle behind this method is the theorem of Pythagoras. That is, in a right-angled triangle the square of the hypotenuse is equal to the sum of the squares of the other sides. And a triangle with sides 3 m, 4 m and 5 m satisfies this theorem. The same holds for a triangle with sides of lengths proportional to these numbers.

3.2.7 Second experimental procedure to find the CG

The first method for finding the center of gravity was described in the previous experiments of balancing circles, parallelograms and triangles horizontally above a vertical bamboo skewer. This is the simplest and most intuitive way to understand the meaning of the center of gravity. With this procedure we can also perceive that it is a single point for each body. Experiment shows that these bodies only remain in equilibrium when supported by a single point called the CG. But there were conceptual problems with this approach, as we saw before. We return to these geometric figures and perform another set of experiments.

We now present the second method for finding the CG of these figures, which avoids the problems already presented. We use plane pasteboard figures of the same shape and size as before. But now we make two or three holes in each figure with nails or a single-hole punch-pliers. The diameters of the holes should be small compared with the dimensions of the figures (so that they will not change the weight or matter distribution of the figures appreciably), but large enough for these figures to hang freely on the pin or hook. That is, the friction between the pin and the figures should be very small, such that the figures can oscillate freely around the pin. Single-hole punch-pliers are very practical and work very well with pasteboard figures with dimensions larger than 5 cm. The circular holes they make allow the figures to swing freely when they hang by a pin or even on a horizontal barbecue bamboo skewer.

Experiment

We make a small hole in a pasteboard circle equal to the one used before. The hole should be made in an arbitrary position which does not coincide with the center of the circle. We then hang this circle on a pin

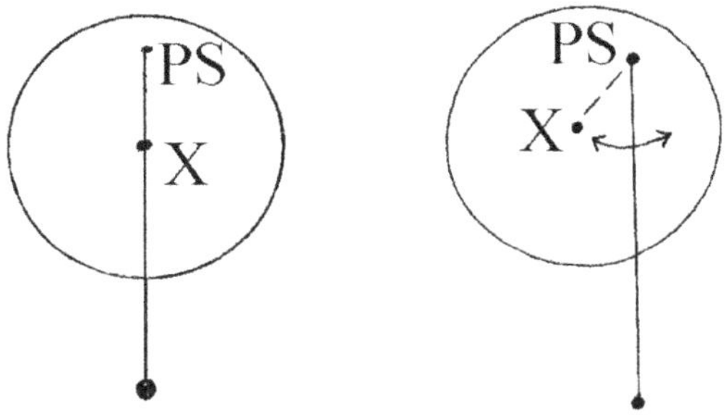

Fig. 3.24. The circle remains at rest when released in the preferential position, otherwise its center will oscillate around the vertical passing through the point of support until it stops in the preferential position due to friction.

stuck into a vertical bamboo skewer. That is, with a horizontal pin the plane of the circle will be vertical. The location of the pin will be represented in the next Figures by the letters PS, indicating that it is the point of suspension. The plumb line is also placed on the pin. We wait until the system reaches equilibrium. Experience shows that in equilibrium the center X of the circle will be vertically below the pin, as indicated by the plumb line, as in Fig. 3.24.

This is also a preferential position in the sense that, if the circle is released from rest from this position, it remains in equilibrium. If the circle is released from rest with its center outside the vertical passing through the pin, we observe that the center of the circle swings around this vertical until it stops in the preferential position due to friction, as in Fig. 3.24.

Instead of hanging the circle on the pin, we could also tie the pierced circle with a thread passing through its hole. The upper end of the thread is then attached to a support above the circle. We again observe the same phenomena as before, provided the circle is free to oscillate around the thread. That is, the downward extended vertical along the thread will pass through the center of the circle when it reaches equilibrium.

We can now present the second experimental procedure for finding the CG.

We consider a pasteboard circle with two or three small holes pierced in arbitrary locations. We hang it with the pin passing through one of its holes, and release the circle from rest. Normally it oscillates and reaches equilibrium. We hang a plumb line by the pin and wait for the whole system to reach equilibrium. We then use a pencil to draw a

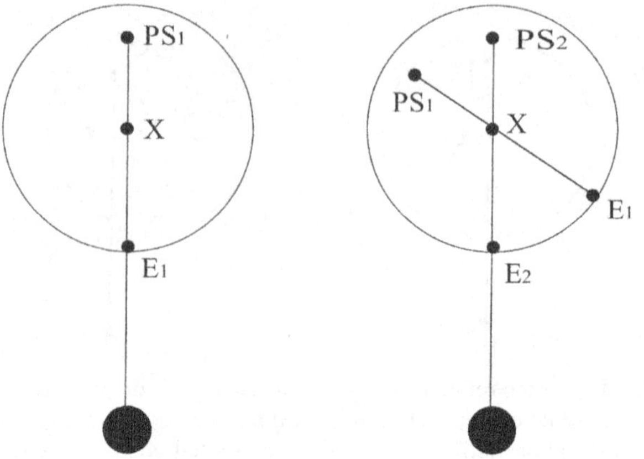

Fig. 3.25. Second experimental procedure to find the CG of a circle.

straight line in the circle coinciding with the vertical indicated by the plumb line. We call it PS_1E_1, where PS_1 is the point of suspension indicated by the pin (these letters should be written at the side of the first hole) and E_1 is the bottom end of the body along this vertical, as in Fig. 3.25.

The plumb line and the circle are removed from the pin. We now repeat the procedure, this time hanging the circle by a second hole PS_2. We hang the plumb line, wait for the system to reach equilibrium and draw a second vertical PS_2E_2, as in Fig. 3.25.

Experience shows that the two straight lines PS_1E_1 and PS_2E_2 intersect at a point which coincides with the center of the circle. If we repeat the procedure by hanging the circle and plumb line by a third hole PS_3, the third vertical PS_3E_3 will also pass through the center of the circle. It is convenient to draw three or more lines like this in order to find the point of intersection with greater precision. This procedure also shows that all verticals intersect at a single point.

But this coincidence of all points of intersection is not always perfect. One reason for this fact is the friction that always exists between the circle and the plumb line while the system is oscillating, before reaching equilibrium. Sometimes this friction prevents the plumb line from reaching a vertical direction when at rest, as the line can stick on irregularities in the pasteboard. But the main reason for a lack of coincidence of all points is the difficulty in drawing the verticals upon the figure to coincide with the plumb line. We have to attach the thread with our fingers in order to draw the lines. At this point we could change the real direction indicated by the plumb line very slightly.

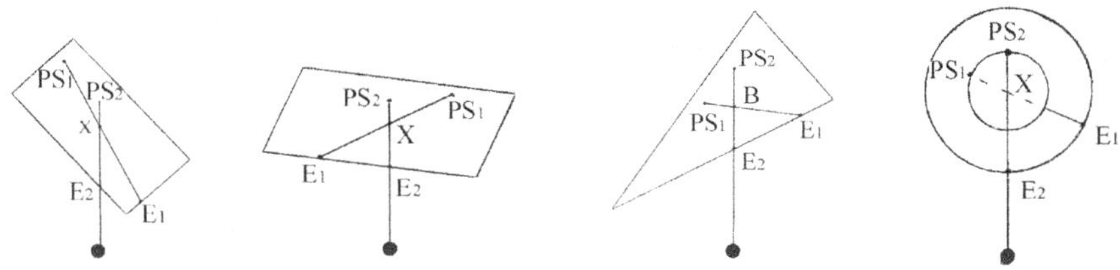

Fig. 3.26. Second experimental procedure to find the CG of a rectangle, of a parallelogram, of a triangle and of a washer.

But with a little practice and patience we can optimize this procedure. Then we can say with certainty that all verticals intersect at the center of the circle. Remember that we are considering holes of small diameters as compared with the size of the figure. This means that these holes do not significantly disturb the weight or the matter distribution of the figure.

Experiment

This procedure is repeated with a rectangle and a parallelogram, by piercing two or three small holes in each figure. The verticals are drawn and we observe that their intersections coincide with the centers of these figures, as in Fig. 3.26.

By repeating the same procedure with a triangle we find that the intersection of the verticals coincides with the barycenter of the triangle, as in Fig. 3.26.

Experiment

We can repeat the procedure with a pasteboard washer by piercing two or three holes in it and hanging it by a pin. Alternatively the washer can be hung by its interior circumference, keeping the washer in a vertical plane. We then hang the plumb line by the pin and draw the first vertical line. By repeating the procedure with another point along the internal circumference, we find that the intersection of the verticals coincides with the center of the washer, as in Fig. 3.26. This agrees with the intersection of the two stretched threads performed before. We can also compare the present experiment with the one in which we used two loose threads. In this case the vertical passing through the intersection of the loose threads coincides with the direction of the vertical bamboo skewer placed below them or with the downward projection passing through the hook holding the threads. That is, this vertical coincides with the axis of symmetry of the washer. And this axis of symmetry also passes through the geometric center of the washer. This means that all verticals drawn in this experiment intersect at the center of the washer.

Experiment

We now repeat this procedure utilizing a pasteboard Moon in first quarter or with a pasteboard letter C. Once more We observe that the intersection of all verticals coincides with the previous experiment performed with stretched threads.

Experiment

We cut out a plane pasteboard figure of arbitrary shape, devoid of any symmetry. Two or three small holes are pierced in the figure. We then localize its CG by the first procedure. That is, we try to find the specific point at which the vertical stand must be placed in order for the figure to remain in equilibrium in a horizontal plane when released from rest. We mark this point with a pen in both sides of the pasteboard. Then we use the second procedure to locate the CG. That is, we hang the figure in a vertical plane by a horizontal pin passing through one of its holes and wait for equilibrium. We then draw a vertical line with the help of a plumb line. We observe that it passes through the CG obtained with the first procedure, although the figure lacks symmetry. The same happens when we hang the figure by the second or third hole.

The essence of these experiments can be stated as follows. A rigid body hangs by a point of suspension PS_1, such that it is free to rotate in all directions around this point. For each PS there will be a preferential position such that the body will remain in equilibrium when released from rest. If it is not let go in this preferential position, when released from rest it will oscillate around the vertical passing through PS, until it stops due to friction. After the body reaches equilibrium, a vertical is drawn passing through PS_1. Choose a second point PS_2 outside this vertical. The body is suspended by PS_2 and the procedure is repeated. Experience shows that the two verticals obtained in this way intersect at a single point. The same happens when the body is suspended by any other point PS. That is, all verticals passing through the points of suspension intersect at a single point.

Practical definition CG6

These facts lead us to a more general definition of the CG:

> **Practical Definition CG6:** The center of gravity of a body is the intersection of all verticals passing through the points of suspension when it is in equilibrium and is free to rotate around these points.

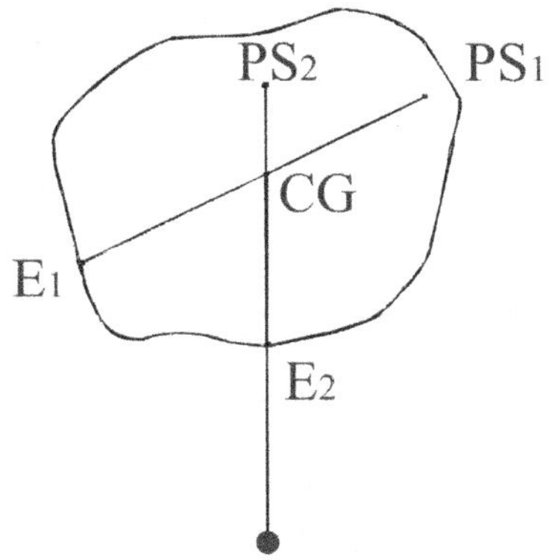

Fig. 3.27. Second experimental procedure to find the CG of a figure with an arbitrary shape.

The detailed procedure or finding the CG by drawing the verticals through the points of suspension has already been presented. It is illustrated in Fig. 3.27 for a body of arbitrary shape.

Experience teaches us that the CG is unique for each body. Moreover, it does not need to coincide with any material part of the body, as we have seen with concave or pierced figures. It is important to emphasize two points in this practical definition. (A) The body must be free to rotate around the point of suspension. We can keep a homogeneous ruler in equilibrium horizontally, for instance, by holding it at one end with our fingers, provided we press our fingers together to prevent the ruler from rotating. In this case we should not draw the vertical line through the point of suspension because the figure is not free to rotate. If we let the ruler oscillate around our fingers, it will not remain in this position when released. Instead it will oscillate, stopping with its larger axis in the vertical direction. (B) We should only draw the verticals in order to find the CG after the body has reached equilibrium, that is, when all its parts are at rest relative to the Earth. No vertical should be drawn while it is oscillating around the equilibrium position.

This last definition of the CG is much more abstract than CG2. Definition CG2 is more intuitive and clearly indicates the existence of a single, specific point in each body, such that it remains in equilibrium under the action of gravity when supported by this point. But definition CG2 has problems when dealing with concave or volumetric bodies, as we saw before. Definition CG6 is more general and can be applied to all cases considered here.

A three-dimensional body must be suspended by a thread attached to one of its external points PS_1. We wait until the body reaches equilibrium. Then we must imagine the vertical extended downward through PS_1 until it reaches the end E_1 of the body. We then suspend the body by

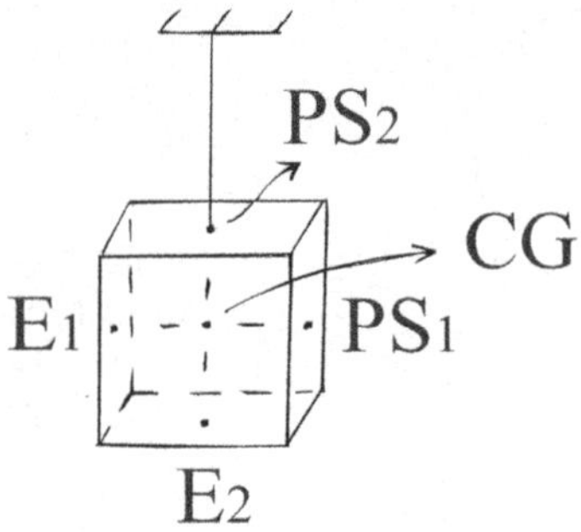

Fig. 3.28. Finding the CG of a cube by the second experimental procedure.

the thread attached to another external point PS_2. We wait until the body reaches equilibrium and imagine the vertical extended downward through PS_2, reaching another external point E_2 of the body. The intersection of these two verticals is the CG of the body. This procedure is illustrated for the case of a cube in Fig. 3.28.

Now that we have a clear and general practical definition of the CG, we can clarify the concepts related to the support and suspension of a body.

Definition: We say that a body in equilibrium is supported by a point (or by a small region or surface) when this point of support is below the CG of the body. This aid or support point will be represented by the letters PA.

Definition: We say that a body in equilibrium is suspended by a point (or by a small region or surface) when this point of suspension is above the CG of the body. This point of suspension will be represented by the letters PS.

3.2.8 Third experimental procedure to find the CG

We now analyze the experiments performed earlier with three-dimensional bodies. The cube or die remained in equilibrium when the vertical bamboo skewer was placed under the center of each one of its sides. By extending these six verticals upward from the support point PA (the center of each face), we find that they intersect at the center of the cube. The same happens with the verticals extended upward from the centers of the six external faces of the screw-nut: they intersect at the center of symmetry of the nut. The sphere remains in equilibrium when supported by any point on a flat table. The verticals extended upwards from these points of support all meet at the center of the sphere. The egg

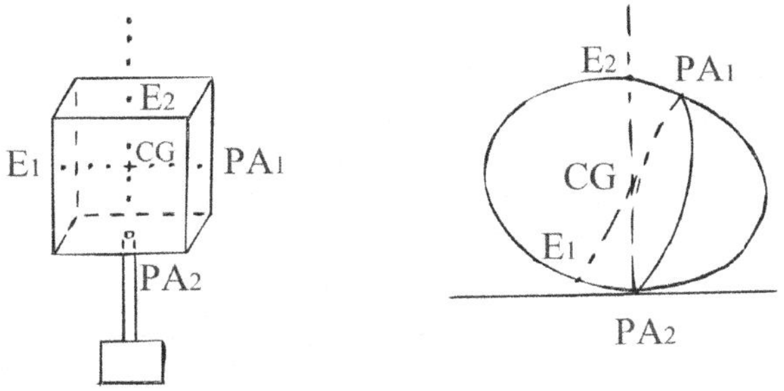

Fig. 3.29. Finding the CG of a cube and of an egg by the third experimental procedure.

remained in equilibrium on a horizontal table when supported by any point along a specific circumference of its shell. By supporting the egg by two or three of these points belonging to this specific circumference and extending the verticals upward through these points of support, we can see that they all meet at a specific point inside the egg.

We first support the body by a point of support PA_1. We extend the vertical passing through PA_1 upward to E_1, where E_1 is the upper end of the body along this vertical line. We then support the body by another point of support PA_2 which is not along the first vertical line. We extend the second vertical passing through PA_2 upward to E_2, where E_2 is the upper end of the body along this second vertical line. The intersection of these two verticals is the CG of the body, as in Fig. 3.29.

That is, it is possible to locate the CG of a body both by the intersection of the downward verticals drawn from the points of suspension, and by the upward verticals drawn from the points of support.

Practical definition CG7

This suggests another practical way of finding the CG:

> **Practical Definition CG7:** The center of gravity of a body is the intersection of the verticals extended upwards from the points of support when the body is in equilibrium and is free to rotate around these points.

The center of gravity obtained by practical definition CG6 always coincides with the center of gravity obtained by practical definition CG7. This can be seen, for instance, by hanging any of these three-dimensional bodies by threads connected to a rigid support. The thread can be tied to the bodies if they have holes, or attached to them with chewing-gum or with a piece of modeling clay.

Let us suppose, for instance, that we tie the upper end of a thread to a fixed support and attach the lower end to a sphere with chewing-gum. We release the system and wait until it reaches equilibrium. In this case the point of suspension (where the chewing-gum touches the sphere) will be vertically above the center of the sphere. The same is true for other bodies.

3.2.9 Conditions of equilibrium for supported bodies

We now conclude this initial section with a few more experiments. They are very simple but extremely important. We will work with bodies for which the centers of gravity have already been determined experimentally. Some of these new experiments (or parts of them) were performed above. Here we will establish the conditions of equilibrium and motion for bodies supported from below, that is, for which the CG is above the PA.

Experiment

We will work with a triangle, but the experiment can be performed with any plane figure for which the CG coincides with one of its material points. We first use a pen to mark the CG (barycenter) of the triangle. We then try to balance it horizontally by placing it on several supports and releasing it from rest. We begin with a vertical bottle. Equilibrium occurs whenever the CG of the triangle is above the bottle cap. If the vertical through the CG falls outside the bottle cap, the triangle falls down, its CG approaching the surface of the Earth. Next we use a vertical pencil placed standing on its tip inside a sharpener. Once more, equilibrium occurs only when the CG of the triangle is above the horizontal end of the pencil. We now utilize a vertical bamboo skewer with its tip stuck in a clump of modeling clay. Once more we can balance the triangle horizontally as before, but there is not much freedom left here. That is, any small horizontal motion of the CG which removes it from the upper end of the bamboo skewer makes the triangle fall to the ground. When we use a vertical bamboo skewer with its tip pointed upward as a stand, it is very difficult to balance the triangle. Any shaking of our hands when we release the triangle is enough to unbalance it and cause it to fall. The same happens with any leaning or quivering of the bamboo skewer due to wind or some other factor. Finally, it is extremely difficult to balance the triangle on the tip of a vertical pin or needle. Sometimes we can only succeed if we stick the pin in the pasteboard (finishing with the experiment) or deform the triangle a little. Many people never succeed in balancing the triangle horizontally on the tip of a vertical needle, no matter how long they try.

Other examples of this fact can be found in one of the previous experiments in which a cube or a metal screw-nut was balanced on a vertical bamboo skewer with its tip downward. Equilibrium was achieved only when its CG (the center of symmetry of the cube or nut) was placed vertically above the upper horizontal surface of the bamboo skewer.

We conclude that a body can only remain in equilibrium if its CG is vertically above the region of support. Moreover, it is extremely difficult to balance a body when its CG is vertically above the support in cases where the area of upper end of the support tends to zero, approaching a mathematical point. This can be shown clearly in the next experiment.

Experiment

We make a small circular hole in the pasteboard triangle of the previous experiment. We hang it on a pin stuck in a vertical bamboo skewer. The horizontal pin passes through the hole and the plane of the triangle is vertical. We turn the triangle in such a way that its CG and the pin are aligned vertically, with the CG above the pin. We release the triangle from rest, holding the base of the bamboo skewer firmly. Experience shows that the triangle does not remain in this position. Its CG begins to swing widely around the vertical extended downward through the pin, until the triangle reaches equilibrium, as in Fig. 3.30. In the final position the pin and the CG are vertical, but with the CG below the pin.

Experiment

We now consider a homogeneous sphere on a horizontal table. We can release it from rest in any position, and it remains in equilibrium. If we give it a small horizontal motion, it rolls until it stops due to friction.

Experiment

An analogous experiment can be performed with any cylindrical homogeneous container with its CG along the axis of symmetry (a cylindrical metal can or plastic bottle, for instance). It remains in equilibrium when released from rest in any position. If it is given a small horizontal motion so that it begins to roll around the line of support, it moves until it stops due to friction.

We now perform a series of three experiments analogous to what we did with the egg earlier, but now with a slightly different symmetry which shows more clearly what is happening. We will deal with a cylindrical shampoo bottle with an elliptical cross section (for which b is half

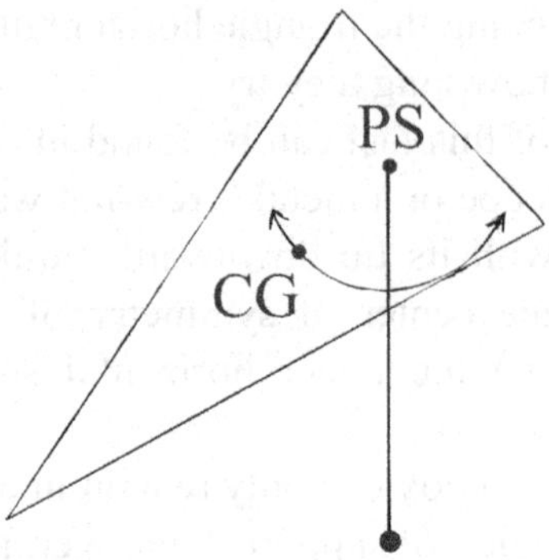

Fig. 3.30. The barycenter of a triangle oscillates around the vertical passing through PS when the triangle is released from rest.

the large diameter or major axis and a is half the small diameter or minor axis, with $b > a$). The center of gravity is along the axis of symmetry of the bottle, passing through the center of the two elliptical bases.

Experiment
The shampoo bottle is set down on a horizontal surface and released from rest. We observe that it only remains in equilibrium when released in such a way that the line of support is along the end of the minor axis 2a, as in Fig. 3.31. In this position the CG is vertically above this line of support. By definition we will call this configuration the preferential position of the vessel.

Experiment
If we turn the vessel slightly around this line and release it, it does not remain at rest. Instead, the straight line connecting the centers of the ellipses will begin to oscillate around the previous vertical line, as shown in Fig. 3.31, until the container reaches equilibrium after stopping due to friction. The final position it reaches is the preferential position. This experiment is analogous to what happens with a rocking chair.

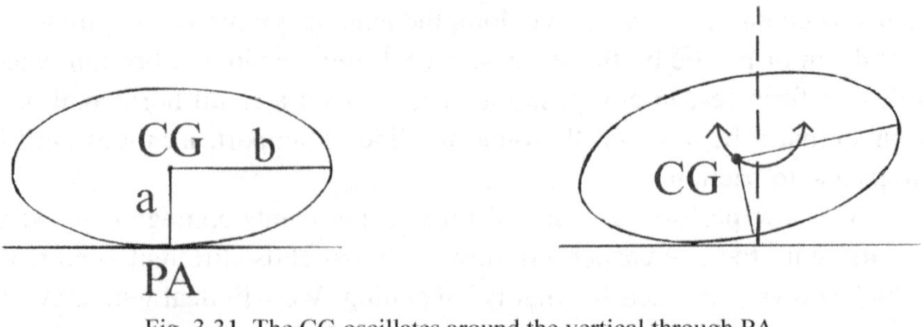

Fig. 3.31. The CG oscillates around the vertical through PA.

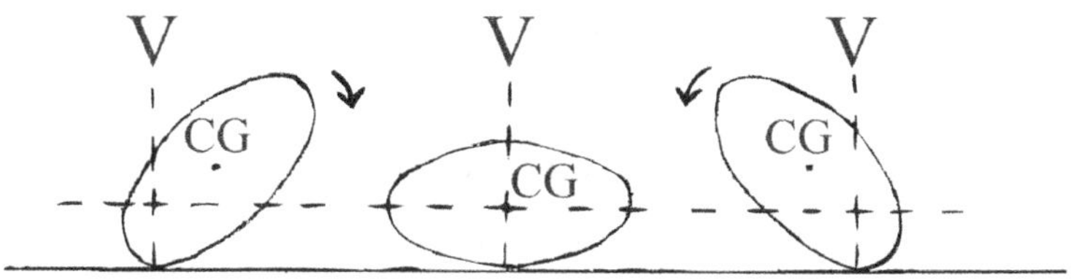

Fig. 3.32. When a body is released from rest the direction of motion is such that the CG moves downwards. The central position is that of stable equilibrium.

We can see in Fig. 3.32 that when we rotate the container around the line below the preferential position, the CG will no longer be along the vertical line passing through the new point or line of contact. Moreover, the CG will be higher in this new position than it was in the preferential position. When the container is released from rest in this new position, the initial direction of motion (that is, the side toward which the vessel will turn) is such that the CG will approach the surface of the Earth. The final position reached by the container, which coincides with the preferential position, is the configuration for which the CG is in the lowest possible position.

Experiment

The container is now released from rest in a position for which the CG is vertically above the lower end of the major axis $2b$. It is practically impossible to balance the container in this position if the floor is flat and smooth. The container always falls toward one or the other side. To find out the side toward which it will fall, we only need to release it from rest with the CG slightly away from the previous vertical line. In this case the initial direction of motion always causes the CG move closer to the ground, as in Fig. 3.33. The final position of equilibrium is once again the preferential configuration.

These and other analogous experiments can be summarized as follows. Suppose a rigid body is placed on flat horizontal surface and released from rest. It will remain in equilibrium only if its CG is vertically above the surface of contact. If the downward projection of the CG lies outside the region of contact, the body will not remain at rest. The initial

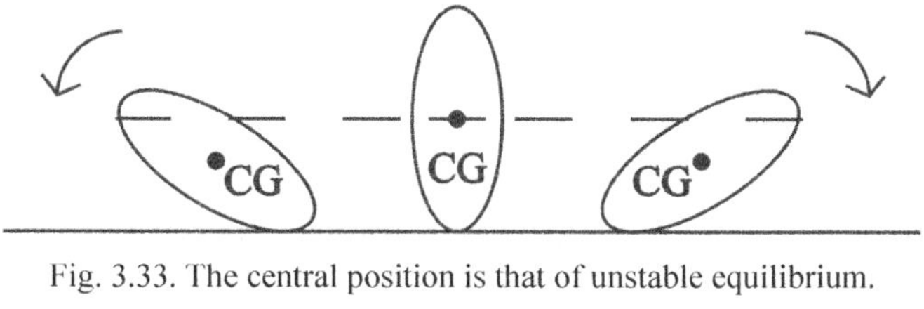

Fig. 3.33. The central position is that of unstable equilibrium.

direction of motion in this case is such that its CG will approach the ground.

3.2.10 Definitions of stable, unstable and neutral equilibrium

These experiments suggest the following definitions.

Stable equilibrium:
This occurs when the CG is vertically above the region of contact and, moreover, any perturbation in the position of the body increases the height of its CG. We call this configuration the preferential position of the body. It is observed experimentally in these cases that any perturbation in the body will cause the CG to oscillate around the vertical passing through the region of support in the preferential configuration, with the body swinging until it reaches equilibrium, because friction will decrease the amplitude of oscillation. In the final position it returns to the initial configuration of stable equilibrium.

Neutral equilibrium:
This occurs when the CG is vertically above the region of support and, moreover, any perturbation in the position of the body does not change the height of its CG relative to the ground. In these cases the body remains in equilibrium for any position in which it is released from rest. If the body is given a small push and begins to move, it will continue to move in this direction until it stops due to friction.

Unstable equilibrium:
This occurs when the CG is vertically above the region of support and, moreover, any perturbation in the position of the body decreases the height of its CG relative to the ground. Any perturbation in the position of the body will move its CG away from the initial position, and it will not returning to that position.

Yet another property connected with the equilibrium of a body supported from below can be derived from these conditions of stable and unstable equilibrium. This property can also be verified experimentally. To do so, we use a rectangular parallelepiped of sides a, *b* and *c*. It can be a brick, a homogenous wood block, a match or shoe box, *etc.* We will always work with the surface bc in a vertical position. From symmetry considerations, and also experimentally, it is easy to verify that the CG of the homogeneous parallelepiped is located at its center. We place a plumb line at the center of the face bc. If the body is a homogenous wood block, the simplest procedure is to put a nail at the center of the surface and tie the thread attached to a plumb onto it. If the parallelepi-

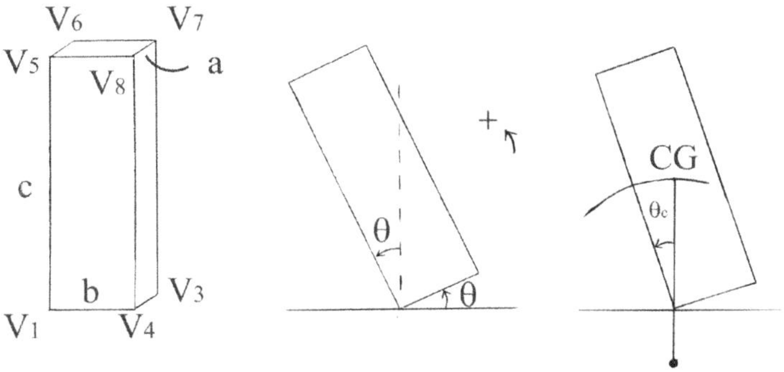

Fig. 3.34. A brick, rotation of an angle θ and the critical angle θ_c for which the CG is in its highest position.

ped is a shoe box, we can pass a bamboo skewer through the centers of both parallel faces of sides *b* and *c*. We then tie a plumb line onto it. For a match box we can pass a pin or needle through the centers of both faces, and then tie a plumb line to it. To prevent the parallelepiped from falling to the ground due to the weight of the plumb line, it is important for the weight of the plumb line to be much smaller than the weight of the parallelepiped. The experiment does not work as well if the parallelepiped is very thin, that is, if side *a* is much smaller than sides *b* and *c* (as is the case with a pasteboard rectangle, where the thickness of the rectangle is much smaller than its sides). In these cases it is difficult to balance the body with surface *bc* in a vertical plane. After everything has been prepared we begin the experiments.

Experiment

We begin with the parallelepiped at rest above a horizontal table, with side *c* vertical and side *b* horizontal. Surface *ab* is horizontal, together with its four vertices V_1, V_2, V_3 and V_4, as in Fig. 3.34.

We define rotation in the vertical plane around the horizontal axis V_1V_2 when V_5V_6 moves down and V_3V_4 moves up as indicating a positive angle, as in Fig. 3.34.

If we rotate the parallelepiped around the axis V_1V_2 of an angle θ and release it from rest, its initial motion is such that its CG falls, as we saw in the conditions for stable and unstable equilibrium. It is easy to see that there will be a critical angle θ_c for which the straight line passing through V_1V_2 and by the CG will be vertical, coinciding with the direction of the plumb line, as in Fig. 3.34. In this situation the CG is in its highest position. If the parallelepiped is released at rest from an initial angle smaller than the critical angle, it will tend to return to the position with side *c* vertical and side *b* horizontal, $\theta = 0°$, as a rotation in this

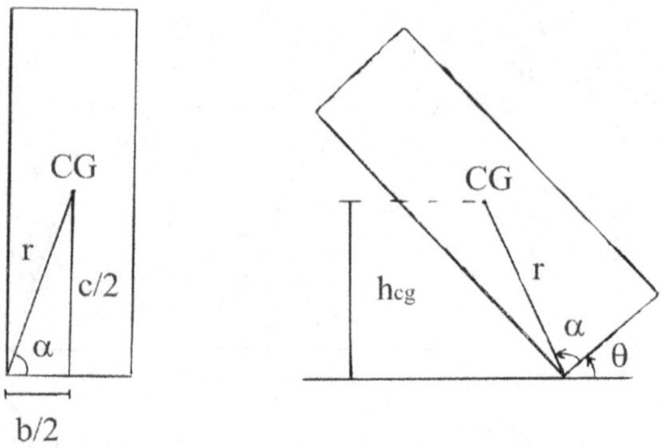

Fig. 3.35. Geometrical properties of a brick.

sense will lower the CG. If the initial angle is higher than the critical angle, the body will tend to move away from the initial position, falling toward the side where c tends to a horizontal position and b tends to a vertical position, $\theta = 90°$. The position of the critical angle is always unstable equilibrium.

From Fig. 3.35 we can see that the tangent of the angle α between the base V_1V_4 and the straight line connecting the vertex V_1 to the CG is given by c/b, $\tan\alpha = c/b$. The magnitude r is given by $r = (c^2 + b^2)^{1/2}/2$.

From Figs. 3.34 and 3.35 we can see that the critical angle θ_c is given by $90° - \alpha$. This means that $\tan\alpha = \tan(90° - \theta_c) = c/b$.

From Fig. 3.35 we can see that in general the value of the height of the CG is given by $h_{CG} = r\sin(\alpha + \theta)$. When $\theta = 0°$ we have $h_{CG} = c/2$, when $\theta = 90°$ we have $h_{CG} = b/2$. The highest value acquired by the CG relative to the ground happens for $\alpha + \theta = 90°$, when $h_{CG} = r = (c^2 + b^2)^{1/2}/2$.

When $c = b$ we have $\alpha = \theta_c = 45°$. In this case the smallest value for the height of the CG is given by $h_{CG} = b/2 = c/2 = 0.5c$. The highest value is given by $h_{CG} = 2^{1/2}c/2 = 0.71c$. If $c = 3b$, $\alpha = 71.6°$ and $\theta_c = 18.4°$. In this case we have $h_{CG} = c/2 = 0.50c$ when $\theta = 0°$, $h_{CG} = 10^{1/2}c/6 = 0.53c$ when $\theta = \theta_c$, and $h_{CG} = c/6 = 0.17c$ when $\theta = 90°$. In the case for which $c = b/3$ we have $\alpha = 18.4°$, $\theta_c = 71.6°$, $h_{CG} = c/2 = 0.50c$ when $\theta = 0°$, $h_{CG} = 10^{1/2}c/2 = 1.58c$ when $\theta = \theta_c$ and $h_{CG} = 3c/2 = 1.5c$ when $\theta = 90°$.

From these conditions we see that the stability of a body supported from below in stable equilibrium increases when the height of its CG decreases. That is, the critical angle increases when we decrease the height of the CG.

We can control this experiment by working with a body of the same weight and external shape, but for which we can change the position of its CG. The idea here is to use a hollow rectangular box of sides a, b and c which has the CG at the center of the box. We will suppose that the side bc is always vertical. We then place another body inside the box, suspended at a height h from the base, as in Fig. 3.35.

What is important now is that we can control this height. For a match box, for instance, we can attach a number of fishing sinkers with modeling clay to the lower or upper part of the box. We can then check that the CG of the system box-sinkers is located at some intermediate point between the center of the box and the center of the sinkers. Let us suppose that it is at a height h_{CG} from the base of the box over a horizontal surface, situated along the axis of symmetry of the lower base b of the box, as in Fig. 3.35.

Experiment

We place sinkers inside a match box, along the bottom side, and place the match box on a horizontal surface. We rotate the system around one of the axes of the base, releasing it from rest. We observe that for some angles the system returns to the position for which $\theta = 0°$, while for angles greater than a certain critical angle θ_{cI} the box falls to the other side, towards $\theta = 90°$, moving away from the initial position. We now invert the position of the shots, in such a way that they remain attached internally to the match box, but on its top side. We repeat the same procedure, and now obtain another critical angle θ_{cS}. It is found experimentally that this new critical angle is much smaller than the previous critical angle, or, $\theta_{cS} < \theta_{cI}$.

By the previous definition of equilibrium we conclude that the match box is in a position of stable equilibrium whether the sinkers are below or above. The reason for this is that any small perturbation of this position, for clockwise and for anti-clockwise rotation with initial angles smaller than the critical angle, the box returns to the initial position when released from rest. Nevertheless, we can say that the box with the sinkers on the bottom is more stable than the box with the sinkers on the top, as the critical angle in the first case is much larger than the critical angle in the second case. The size of this critical angle can then be considered the degree of stability of the system.

We now want to know the value of the critical angle θ_c for this system. When the box rotates around the axis V_1V_2 of an angle θ, as in the previous experiment, it returns to the position for which $\theta = 0°$ when released from rest if $\theta < \theta_c$. If $\theta > \theta_c$, the box does not return to the posi-

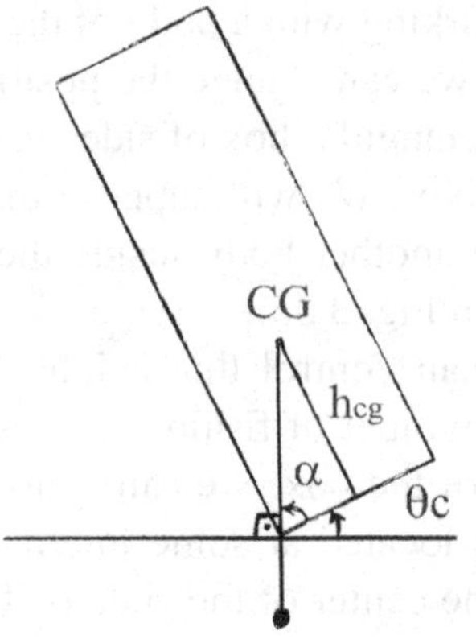

Fig. 3.36. Conditions of stability for a body.

tion for which $\theta = 0°$ when released from rest, but falls to the opposite side, towards $\theta = 90°$. Let α be the angle between the horizontal base b and the straight line connecting V_1V_2 to the CG. We then have $\tan\alpha = h_{CG}/(b/2)$ and $\alpha = \arctan(2h_{CG}/b)$, as in Fig. 3.36.

At the critical angle we have $\alpha + \theta_c = 90°$. That is, $\theta_c = 90° - \arctan(2h_{CG}/b)$. If the height of the CG, h_{CG}, is very small, much smaller than b, the critical angle will be very high, close to 90°, which indicates high stability of the body. If h_{CG} is much larger than b, the critical angle will be very small, close to 0°. Any perturbation in the system will make it fall, moving away from the initial position. From this last equation we conclude that to increase the stability of a system we must decrease the ratio h_{CG}/b. There are two basic ways to do this: (A) Decrease the height of the CG (as we saw for the match box with the sinkers in the lower side), and (B) Increase the base around which the system rotates.

We can then define the stability of a system by its critical angle: the larger this critical angle, the greater the stability of the system. That is, for two systems in stable equilibrium, the system which has the greatest critical angle will be said to have greater stability.

3.2.11 Conditions of equilibrium for suspended bodies

We now consider the main conditions of equilibrium and motion for bodies suspended from above. That is, with the point of suspension PS above the CG of the body. We will consider convex bodies or bodies pierced so that they can hang from a pin passing through a hole or by a thread tied to a hole. Once again we will consider bodies whose centers of gravity we determined earlier, and those where the hole does not coincide with the CG. Some of these new experiments, or parts of them, have already been performed. But they will be repeated in order to

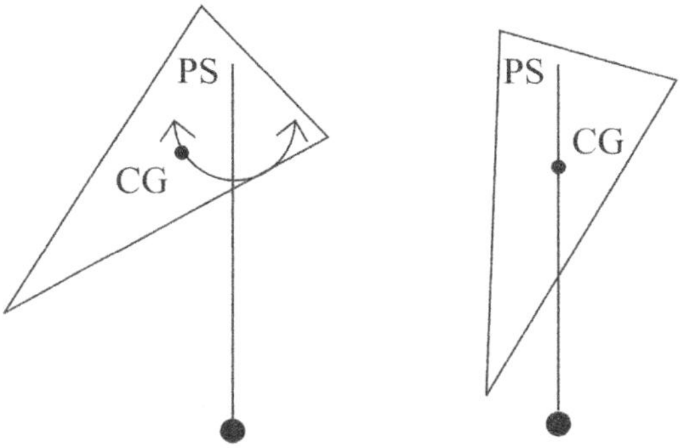

Fig. 3.37. When released the CG oscillates around the vertical passing through PS until the body stops with the CG vertically below the PS.

clearly establish the conditions of equilibrium and motion of suspended bodies. We will work with a triangle, but similar experiments can be performed with other bodies.

Experiment

We hang a triangle by a pin passing through one of its holes, releasing it from rest. We observe that it only remains in equilibrium if its CG is vertically below the PS. This configuration is called the preferential position of the suspended body.

Experiment

We now rotate the triangle around the PS a certain angle, such that the CG and the pin are no longer along a vertical line. The triangle is released from rest. We observe that the CG oscillates around the vertical passing through the PS, as shown in Fig. 3.37. The amplitude of oscillation decreases due to friction. When the triangle stops, it returns to the preferential position with the PS and CG along a vertical line. Moreover, in equilibrium the CG is below the PS.

From this Figure we can see that the preferential position has the CG in the lowest position. Any clockwise or anti-clockwise rotation of the triangle around the PS increases the height of the CG.

Experiment

We begin with a symmetrical bicycle wheel (that is, one with the CG at the center of the wheel), at rest, suspended by a horizontal axis. The wheel is attached to the axis by a ball-bearing, in such a way that it is not loose on the axis. We could also use a pasteboard disc pierced at the center. We pass a nail through the center of the disc, with a diameter a little smaller than the diameter of the hole, in such a way as to leave lit-

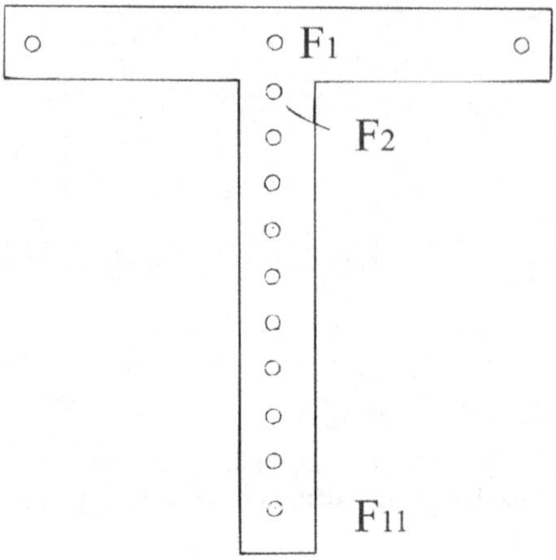

Fig. 3.38. A pasteboard T with holes.

tle room between them, only enough space for the disc to rotate around the axis. The plane of the disc or bicycle wheel should be vertical, with the axis horizontal. We can release the wheel or disc from rest in any position and it remains in equilibrium. If we rotate the wheel or disc slowly one way, giving it a small angular rotation, it continues to rotate in this sense, its angular velocity decreasing due to friction, until it stops.

In these cases the wheel and the disc are suspended by the upper part of the axis, which is a little above the CG of the bodies (located at the center of the wheel or disc). Nevertheless, any rotation of the wheel or disc around the axis does not change the height of the CG relative to the ground.

These experiments suggest the following definitions:

Stable equilibrium:

When the CG is vertically below the PS and, moreover, when any perturbation of this position moves the CG upwards. If this happens, experience shows that the body returns to the preferential configuration.

Neutral equilibrium:

This occurs when the CG is vertically below the PS and when any perturbation in this position does not change the height of the CG relative to the ground. In this case the body remains in equilibrium in any position where it is released from rest. If it receives a small impulse and begins to rotate around the PS, it will continue to turn in this direction until it stops due to friction.

Before we move on, another experiment is worth performing. We cut out a pasteboard figure in the shape of the letter T. The length from the tip of one arm of the T to the tip of the other arm should be 15 cm. The height of the T should be 15 cm or 20 cm. The width of the arms and body of the T should be 2 cm. We pierce 10 or 11 holes along the axis of symmetry of the T. We call them F_1 to F_{11}, with F_1 at the intersection of the arms and with F_{11} at the end of the figure. We can also make holes at the hands of the two arms, as in Fig. 3.38.

Experiment

To begin with, we locate the CG of the T. This can be done, for instance, by hanging it by the holes at the hands and drawing the verticals in equilibrium. The CG is the intersection of these two verticals, which should be along the axis of symmetry of the T, closer to F_1 than to F_{11}. The T will then be released from rest, hung by a hole along its axis of symmetry, with the arms horizontal and its body below the arms (that is, with F_1 above F_{11}). When it is suspended by holes above the CG, such as F_1 or F_2, it remains in equilibrium at the position in which it was released. On the other hand, when it is released from rest by holes below the CG, such as F_{10} or F_{11}, it turns to one side or the other, swings a few times with decreasing amplitude, and stops with F_{11} above F_1. This experiment again illustrates that the configuration with the CG vertically above the PS is unstable if the area of support is very small, like a point. On the other hand, the configuration with the CG vertically below the PS is stable, even when the area of support is very small, like a point. Although the explanation for this experiment is based on principles we have already seen, it is very interesting. After all, all holes have the same diameter and allow the same rotation around the PS. But only in certain cases will the body rotate when released from rest, inverting the position of the arms relative to the body of the T.

3.2.12 Definitive definition of the center of gravity

Cases in which the CG coincides with the PS

It may be impossible to do a real experiment in which a body is suspended or supported by a point which passes exactly through its CG, and is free to rotate around this point. Even when we try to approximate this situation from below, the CG will be always slightly above the point of support PA. This is the case, for instance, where the pasteboard triangle is horizontal and supported by a vertical base placed below the barycenter of the triangle. Here the point of contact between the base and the

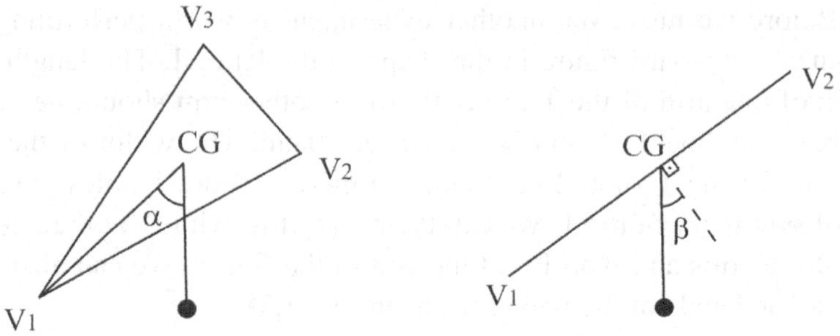

Fig. 3.39. A body suspended exactly by the CG and free to rotate around this point does not move when released from rest, no matter its orientation relative to the Earth.

pasteboard is slightly below the CG of the triangle, which is located at a point at the middle of the pasteboard. When we try to approximate this situation from above, the CG will always be slightly below the PS. This is the case, for instance, of a triangle in the vertical plane suspended by a horizontal pin passing through a hole made at the barycenter of the triangle. The diameter of the hole must be slightly larger than the diameter of the pin, in order to allow free rotation of the triangle. In this case the PS will be the point of contact between the pin and the upper part of the hole, while the CG will be at the center of the hole.

Another difficulty arises for three-dimensional bodies. For instance, if we have a solid parallelepiped, we can only support it by a stick touching its outside surface, or by a thread tied to its outside surface. Its CG is located in the middle of the parallelepiped, inside the brick. In order to suspend or support it by this point we need to make a hole in the parallelepiped; and this would change its distribution of matter. But if the width of this hole is very small compared with the sides of the parallelepiped, we can neglect this modification to the matter of the brick.

From what we have already seen, we can imagine what would happen if we could perform this experiment. We have already seen that the CG of any rigid body released from rest tends to move toward the surface of the Earth. If the body is suspended exactly by the CG, being free to rotate around this point, any rotation of the body will not change the height of the CG relative to the ground. In this case the body would remain in equilibrium for all positions in which it was released, no matter what its orientation relative to the ground.

Let us consider an idealized experiment with a triangle in a horizontal plane suspended from above exactly by its barycenter by a vertical string, or supported by a vertical stick under it. If released from rest the triangle will remain in this position regardless of the orientation of any of its vertices relative to ambient objects (*e.g.*, pointing toward a tree or a house, or pointing toward North or West).

Let us now suppose that the triangle is released from rest in a vertical plane with a horizontal axis orthogonal to its plane passing exactly through its CG and attached to the triangle. This axis is supported in both sides around the triangle by two vertical stands. The triangle is considered free to turn in any direction around this axis. After release the triangle will remain in its initial position regardless of the orientation of any of its vertices relative to a vertical passing through the CG.. Let us call V_1 one of its vertices and α the angle between the segment connecting V_1 and the CG with the vertical passing through the CG, as in Fig. 3.39. In other words, the vertical triangle supported by the CG and free to rotate around an axis passing through this point would remain at rest after release, regardless of the value of α. We can approximate this situation experimentally by passing a needle or toothpick through the barycenter of a triangle, orthogonally to its plane. It is difficult to pass the needle exactly through the CG. This means that in general there will be a preferential position for the stationary situation with the CG of the triangle below the needle. But we can then place a very small amount of modelling clay on the upper edge of the triangle vertically above the needle in order to adjust the CG of the whole system to coincide approximately with the needle's axis. In this case we can obtain a result like that of Fig. 3.39 in which the triangle will remain at rest regardless of the angle α of release.

Let us now consider an axis in the plane of the triangle and passing through its CG, rigidly connected to the triangle. We can imagine it extended on both sides beyond the edges of the triangle and supported horizontally at both ends by two rigid vertical stands placed under them. The plane of the triangle is free to rotate in any direction around this axis. The rigid triangle defines a plane. We consider a straight line orthogonal to the plane of the triangle and passing through its CG.. We call β the angle between this straight line and the vertical line passing through the CG, as in Fig. 3.39. If released from rest and supported by the CG, free to rotate around the horizontal axis passing through this point, the triangle should remain at rest in this position regardless of the value of β.

We have seen from the previous real experiments that the CG of any rigid body tends to move closer to the surface of the Earth when it is released from rest. This means that if a body hung exactly by its CG were released from rest, being free to rotate about this point, it would

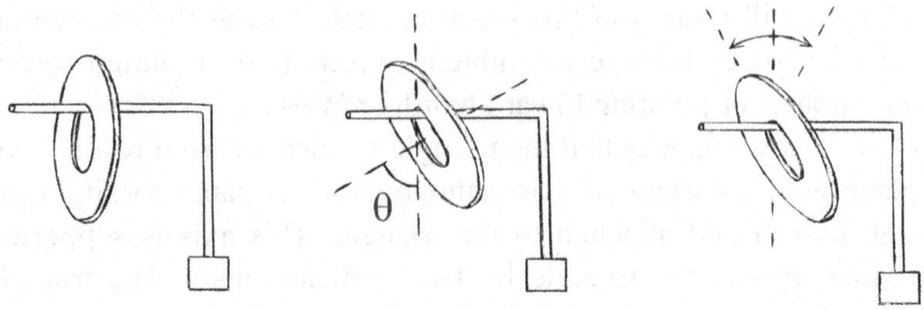

Fig. 3.40. A washer can remain at rest when supported by its internal circumference. However, it does not remain in equilibrium for all orientations of release. If $\theta \neq 90°$ its center will oscillate around the vertical passing through the point of support after release.

not move, no matter what its initial position or orientation relative to the ground. After all, if it did begin to rotate in any direction, the CG would remain at the same height. As the tendencies to rotate in opposite directions cancel one another, it will not begin to rotate when released from rest. This leads to a new and definitive definition for the CG.

Definitive Definition CG8: The center of gravity of any rigid body is a point such that, if the body is suspended from that point, and is released from rest and free to rotate in all directions around this point, the body so suspended will remain at rest and preserve its original position, no matter what the initial orientation of the body relative to the ground.

If this point is located in empty space, as for concave or pierced figures, we have to imagine a rigid structure connecting the body to this point, so that the body can be suspended from the point.

Later on we will see that Archimedes seems to have defined the CG in this way.

The main difference between definition CG8 and definition CG4 is that in CG8 we say that the body will remain in equilibrium when released from rest no matter what the initial orientation of the body relative to the ground. Let us consider a washer, for example. It can remain in equilibrium when released from rest in a vertical plane, suspended by any point on its internal circumference, as in Fig. 3.40. In this case the axis of the washer makes an angle $\theta = 90°$ with the vertical line. We define the angle θ as the smaller angle between the axis of the washer and the vertical line.

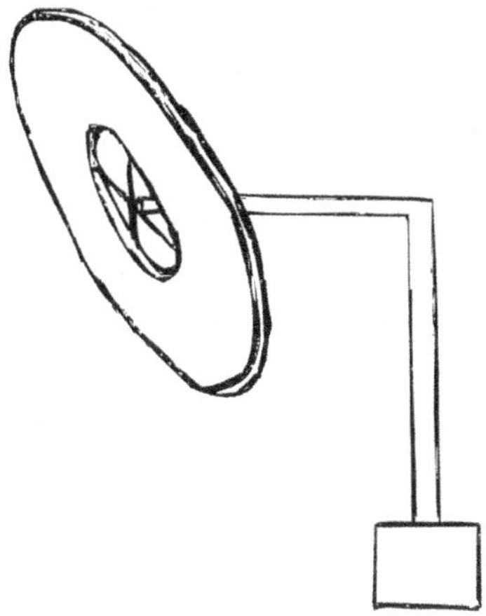

Fig. 3.41. When supported by its CG a body remain in equilibrium no matter the orientation of release.

According to definition CG4, this point of suspension PS along the inside circumference could be considered a center of gravity for the washer. We now suppose that the washer is released from rest when suspended by the same point, but with its axis no longer orthogonal to the vertical line. This means that the body will be released from rest with θ different from 90°. In this case the washer does not remain in equilibrium. After release the plane of the washer will oscillate around the vertical line, until it stops due to friction. In the final position $\theta = 90°$, as in Fig. 3.40.

According to definition CG8, this point of support along the internal circumference cannot be considered the CG of the washer. We have already seen in the practical procedure given by CG6 that the real CG of the washer is its center of symmetry at the center of the washer. When the washer hangs by a point PS located on the inside circumference, the CG will be at its lowest position when it is vertically below this PS, where $\theta = 90°$. This is a position of stable equilibrium. When we decrease the angle θ, the CG moves upward. If it is released from rest in this new position, gravity will cause it to move downward.

Suppose now we attach some spokes in the washer, like the spokes of a bicycle wheel. This can be done with taught threads, or we can consider a real bicycle wheel. Let us suppose the washer or wheel is suspended by its center and is free to rotate in any direction around this point. If it is released from rest with its axis making an angle θ with the vertical line, it remains in equilibrium for any θ, as in Fig. 3.41.

By definition CG8, we can now conclude that this is the real CG of the washer. The reason why it remains in equilibrium for any value of θ, when suspended by its center, is that the height of the CG is independent of θ. And this is the main property of the neutral equilibrium.

We called definition CG8 as definitive. The word definitive should be understood as between quotation marks. The reason is that it is strictly valid only in regions of uniform gravitational forces. These are the regions in which a certain test body is always under the action of the same gravitational force (in intensity and direction) at all points of the region. This is true for small bodies in the vicinity of the Earth. The gravitational forces acting upon each particle of the test body can be considered parallel to one another, and all vertical.

But there are situations in which this is not valid. Let us present a specific example in which we make several assumptions: (A) The body exerting the gravitational force is like the Earth, but shaped like an apple, with the greatest distance between any two particles of this apple-earth given by d_E; (B) the body under the action of the gravitational force is like the moon, but shaped like a banana, with the greatest distance between any two particles of this banana-moon given by d_M; (C) the distance between an arbitrary particle i of this apple-earth and another arbitrary particle j of the banana-moon being given by $d_{ij} = d_E + d_M + \varepsilon_{ij}$, with $0 < \varepsilon_{ij} << d_E + d_M$. In this case a unique center of gravity will not exist. Depending upon the relative orientation between the banana-moon and the apple-earth, there will be distinct lines of equilibrium. In cases like this, the concept of a center of gravity looses its meaning.

Still, definition CG8 may be used for a test body of small dimensions when compared with the radius of the earth.

3.2.13 Summary

We now present the main conclusions we have reached thus far.

Definitions: Equilibrium of a body is when the body and its parts do not move relative to the Earth. The vertical is the straight line traced by a body in free fall at the surface of the Earth, beginning from rest. The horizontal is any straight line or plane orthogonal to the vertical. The center of gravity of any rigid body is a point such that, if the body were to be suspended from that point, released from rest and free to rotate in all directions around this point, the body so suspended would remain at rest and would preserve its original position, no matter what its initial orientation relative to the ground. The CG can be found in practice by the intersection of all the vertical lines passing through the points of suspension of the body when it remains in equilibrium, and is free to rotate around these points of suspension.

Experimental results: The CG is unique for each rigid body. Free bodies fall to the ground when released from rest. The direction of free fall coincides with the direction of a plumb line in equilibrium. Any body can remain in equilibrium after released from rest, provided it is supported from below with its CG located vertically above the surface of contact. Any body can also remain in equilibrium after released from rest if it is suspended by a point PS around which the body is free to rotate, provided the CG is vertically below the PS. Equilibrium will be stable, unstable, or neutral, respectively, when any perturbation in the position of equilibrium of the body increases, decreases, or does not change the height of the CG relative to the ground. When there is stable equilibrium, any perturbation in the position of the body will cause it to oscillate around the position of equilibrium, with decreasing amplitude of oscillation due to friction, until it stops at the position of stable equilibrium. When there is unstable equilibrium, any perturbation in the position of the body will move it away from this position. The initial direction of motion for the perturbed body released from rest will be such that its CG moves downward from its initial height in the position of unstable equilibrium.

Until now we have only described these facts. We are not explaining the experimental data; we are merely summarizing the main results. We will now utilize these basic experimental facts to explain other phenomena that are more complex, but that can be derived from these observations.

3.3 Fun activities with the equilibrist

One of the most interesting classroom activities utilizes a pasteboard equilibrist. It makes the students assimilate and incorporate all the concepts we have seen thus far. It is also fun, especially if performed with several people simultaneously. The idea is to give a problem to the students and to let them solve it by themselves. The teacher should not tell them how to solve the problem and should not explain the causes of the phenomena observed. Only the sequence of tasks needs to be given. This activity should be performed after the students have performed the main experiments. Each student should prepare his own equipment (bamboo skewer, plumb line, a pasteboard equilibrist, *etc.*), and also perform all the procedures described here. At the end of the activity the students can keep their apparatus.

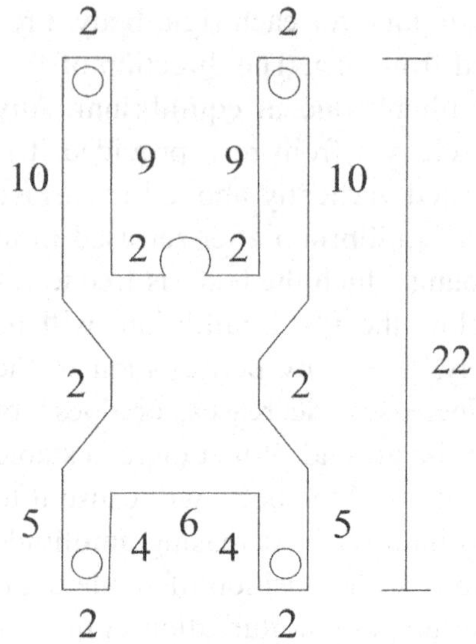

Fig. 3.42. An equilibrist with dimensions in centimeters. There are holes in the hands and feet.

Materials: A support with a plumb line. A pasteboard equilibrist, as in Fig. 3.42, with the dimensions in centimeters. Some modeling clay. Single-hole punch-pliers.

The support with the plumb line could be a barbecue bamboo skewer with the tip stuck in modeling clay, with a pin or needle stuck in the top of the bamboo skewer. The plumb line can be made with sewing thread and a plumb or modeling clay at the bottom, as before. When the equilibrist becomes too heavy with the clay, so that the pin or needle slips out of the bamboo skewer, or the plumb slips off the needle, we can support it with a horizontal bamboo skewer on a table, sticking out from the table, with the plumb line tied to it. In this case the equilibrist will hang by the bamboo skewer itself passing through a hole in the pasteboard, instead of being suspended by the pin, as in the previous case.

The exact dimensions of the equilibrist are not so relevant. The most important for the time being is that it should by symmetric around the body's axis, with the arms pointing up and the legs down, as in Fig. 3.42. The arms should be longer than the legs, as in most situations the equilibrist will be upside down. The dimensions shown in the Figure are appropriate for the activities to be described, in which the paper puppet is balanced by hand by the students.

Another very important property of the equilibrist is that it should be rigid, non-deformable. If we put a large amount of clay on it, a pasteboard equilibrist could bend. In order to prevent this, the pasteboard should be rigid. We can even have an equilibrist made of stiff plastic,

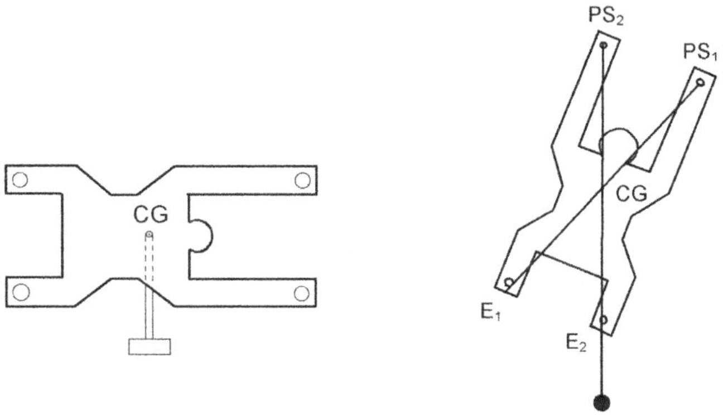

Fig. 3.43. Finding the CG of the equilibrist by the first and second experimental procedures.

which is not so difficult to obtain. If the equilibrist is bent by the clay used in this experiment, what is described will not be observed in some cases.

Initially several identical equilibrists should be cut out from a pasteboard, so that each student receives a figure. The students should pierce the hands and feet of the equilibrist with single-hole punch-pliers, as shown in Fig. 3.42. After this the first task is to locate the CG of the puppet using the two procedures already learnt. (I) We find the point on which the equilibrist should be supported so that it remains in equilibrium horizontally above a vertical stand after released from rest. (II) We suspend it by a needle passing through its two hands, drawing the verticals in each case with the help of a plumb line. The CG should be marked on the pasteboard, preferably on the front and back sides, as in Fig. 3.43.

We then begin with the most interesting part of the game. We ask the students to try to balance the puppet upside down, by placing it above the pointing finger. The finger should be extended horizontally, below the head of the equilibrist. After a few minutes of trials, no one succeeds. Some think that the problem is the curved shape of the head.

We then ask the students to try to balance the puppet with the head upwards and the pointing finger extended horizontally, as if the puppet was sitting on the finger. After several trials, no one succeeds, although now the line of contact is straight and horizontal. For the time being we should not try to explain why they have failed. The idea is to go on with the game.

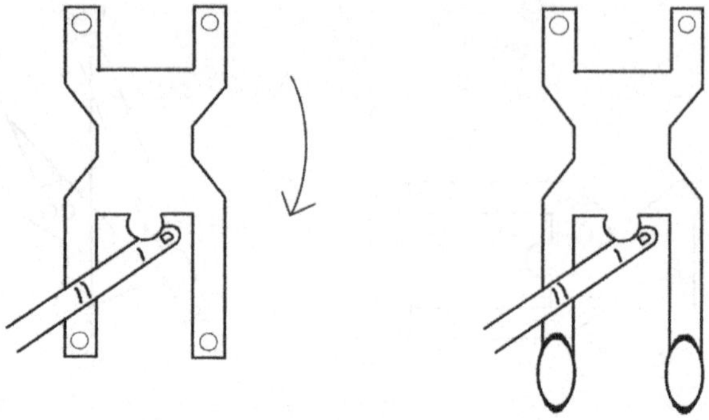

Fig. 3.44. We can keep an equilibrist upside down in our finger by placing enough modeling clay on both of the puppet's hands.

We now ask them to balance the puppet in a horizontal position, placing the pointing finger vertically below it. Now all of them succeed. They easily observe that the puppet's CG is above the finger.

After this we again ask them to balance the puppet horizontally, but now with the pointing finger placed vertically below the head of the puppet. Once more no one succeeds.

Now comes the most stimulating part of the game. We give a piece of modeling clay to each student. We again ask them to try to balance the puppet upside down, on the horizontal pointing finger placed below the puppet's head. We tell them that they can now attach the clay anywhere on the puppet, except on the "hair" of the puppet, that is, on the lower part of the head, to prevent it from sticking to the finger. They can put it on the CG, on the hands, on the legs or wherever they wish. We also tell them that the clay can be attached to the equilibrist as a single

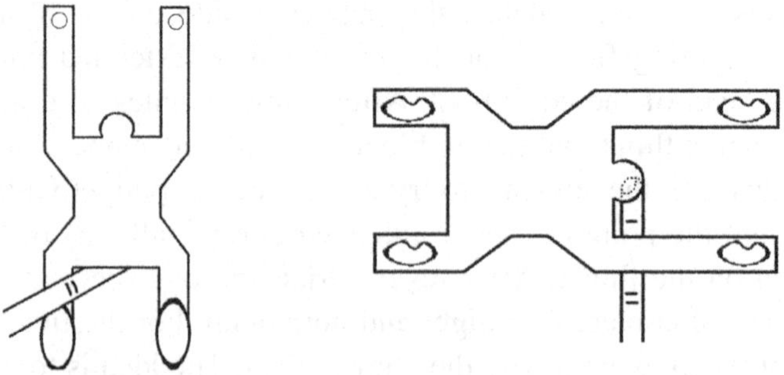

Fig. 3.45. Equilibrating the puppet in a vertical plane with the head on top or in a horizontal plane by placing the pointing finger vertically below the head. In both cases the trick is where to put the modeling clay and how much to use.

whole, or in two or more pieces. The idea here is to encourage the students to experiment, without giving recipes for the solution to the problem. They are shy and leery about what to do at first. But little by little they begin to relax and play the game. After a few minutes one or two students succeed in balancing the puppet upside down on their fingers. The others begin to see what they have done, and in a short time all of them succeed. The secret of success is to place enough clay on both of the puppet's hands, until it remains upside down balanced on our horizontal pointing finger, as in Fig. 3.44.

When an equilibrist does not stay exactly on the vertical, all we have to do is move the clay away from the head (placing it at the tips of the hands, or even hanging from the hands), or increase the amount of clay on the hands. Eventually it will hang vertical and upside down.

After all the students have managed to do the experiment, we ask them to remove the clay and put it somewhere else on the puppet until it stands vertical with the head on top, sitting upon the horizontally extended pointing finger. One or two students will managed this more quickly than before. The others see what they have done and sooner or later all have managed to get the puppet vertical. The secret of success is to place the clay on the feet of the equilibrist, as in Fig. 3.45.

We then ask them to change the position of the clay again until the puppet remains balanced horizontally, supported on the pointing finger extended vertically, placed under the head of the puppet. We ask them to avoid placing clay on the head of the puppet, to prevent it from sticking to the pointing finger. After some effort all of them succeed. (Some students need to see what the others have done before they pick up the trick.) In this case they can attain the final result in several ways, as there is more than one possibility. A common technique is to place clay on the hands and feet in the right amounts until the equilibrist remains horizontal, as in Fig. 3.45.

After this part of the game, we again ask them to change the location of the modeling clay until the puppet remains upside down vertically, supported on the pointing finger extended horizontally under the head of the puppet. Now they will all quickly place enough clay on the hands of the puppet until it reaches the desired position. To show that the equilibrium in this position is very stable, we ask them to rock or blow the equilibrist gently. We can also ask them to balance it over the flat tip of the bamboo skewer, then raise everything with their hands until the arms are stretched. We can even balance the puppet upside down supported on the horizontal needle attached to the bamboo skewer! Even in this case they can rock or blow air on the puppet gently, and it will

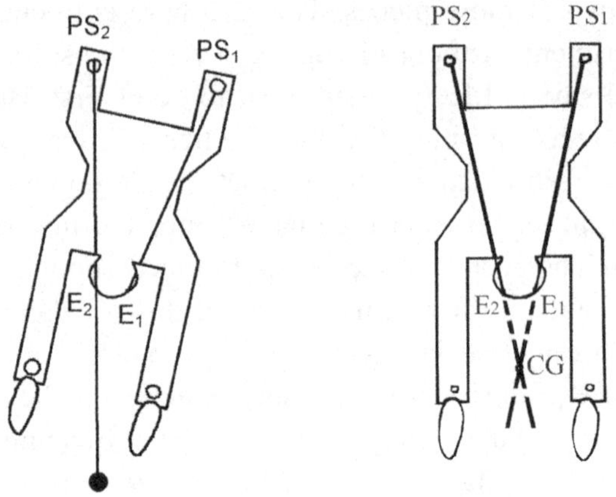

Fig. 3.46. Finding the CG of the equilibrist with modeling clay in both hands.

not fall, but only oscillate around the equilibrium situation, always returning to its vertical position upside down. Everyone admires this. This is a remarkable experience that gives a deep impression. The stability of this equilibrist is really amazing.

We then ask them about the location of the center of gravity in this new situation (equilibrist upside down with clay on the hands). A few of them may think it is located in the same place as before (in the middle of the chest), but the majority will believe that it is located in the head of the puppet, more specifically in its hair. In other words, they believe it is located at the point where the head touches the finger. Without giving the correct answer, we ask them to locate the CG exactly using the second method mentioned above, by suspending the equilibrist (with clay on his hands) through the needle in the support. We first suspend it by the hole in one foot and draw the vertical line with the help of a plumb line. Then we suspend it by the hole in the other foot and draw the second vertical, as in Fig. 3.46. We must tell the students that it is important to locate the CG precisely. This should be done carefully. When they try to do this, some of them say that the method "does not work," because the vertical lines do not seem to intersect (that is, they do not intersect where they expected). Despite this initial reaction, we ask them to continue with the experiment. The final result, when the verticals are carefully drawn, is something like the result shown in Fig. 3.46.

If we extend these two verticals, we see that they intersect outside the head, at a point along the axis of symmetry of the puppet, between the head and the hands (or between the hair and the lower part of the clay).

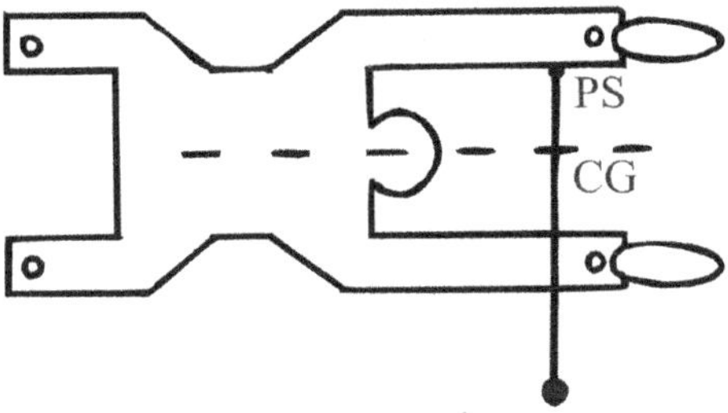

Fig. 3.47. Another way of finding the CG of the equilibrist with modeling clay in both hands.

It is interesting to ask the students to make a drawing like this in their notebooks, full size, utilizing their own puppet with clay on the hands as a model. In order to locate the CG of the puppet with clay on the hands exactly, we ask the students to balance the puppet sideways, in a vertical plane, resting some point of the arm on the horizontal needle, until the body of the puppet stays parallel to the floor. The CG is located at the intersection of the axis of symmetry of the body with the vertical line passing through the needle, obtained with the help of a plumb line, as in Fig. 3.47.

Only after the students have performed all these activities should the teacher explain what has happened. The explanation is that in the cases without clay it was not possible to equilibrate the puppet upside down, nor seated upon the finger with the head at the top, because the CG located at the chest of the equilibrist was always above the point of support PA. And these are conditions of unstable equilibrium. Any shaking of the finger or puppet causes the equilibrist to fall to the ground, because the tendency of the CG is always to approach the surface of the Earth, as in Fig. 3.48. By the same token, it was not possible to balance the puppet horizontally with the vertical finger under its head, because there was no support below the CG located in the middle of the chest. Therefore, when the puppet was released, the CG always fell to the

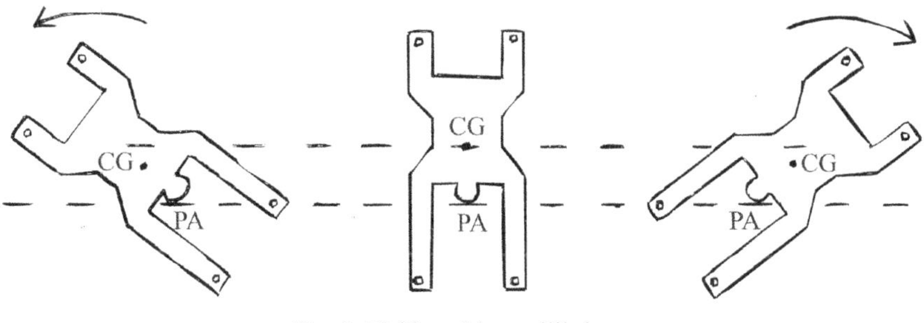

Fig. 3.48. Unstable equilibrium.

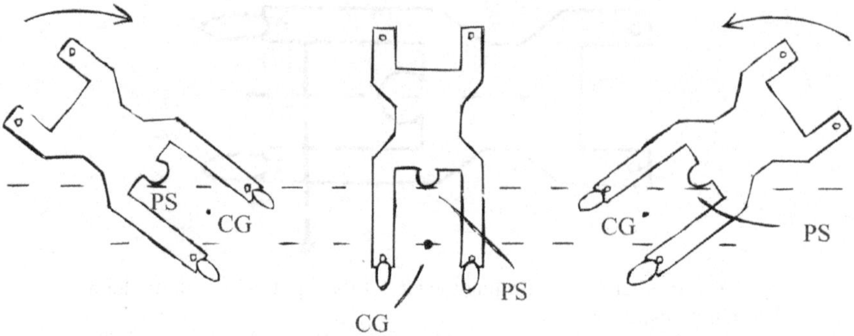

Fig. 3.49. Stable equilibrium.

ground.

On the other hand, when clay was placed on the hands of the puppet and it was balanced upside down on a finger placed under its head, the CG was located below our finger. That is, below the point of suspension, PS. This is a position of stable equilibrium. If we turn the puppet clockwise or anti-clockwise, we raise the CG relative above its height in the position of equilibrium, as in Fig. 3.49.

The same happens when we lean the puppet forward or backward, that is, with the nose or the back of the neck approaching the ground. Again we are raising the CG. This means that any rotation of the puppet around the point of suspension PS increases the height of the CG. As the tendency of the CG is always to fall due to gravity, the puppet tends to return to the position of stable equilibrium after it is released. In this upside down configuration the CG is in its lowest possible position.

When the puppet is sitting on our finger with modeling clay on the feet, the CG is again located between the bottom of the clay and the

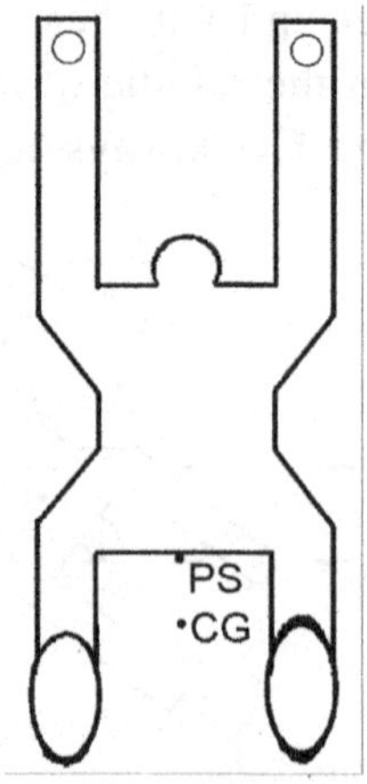

Fig. 3.50. Another situation of stable equilibrium.

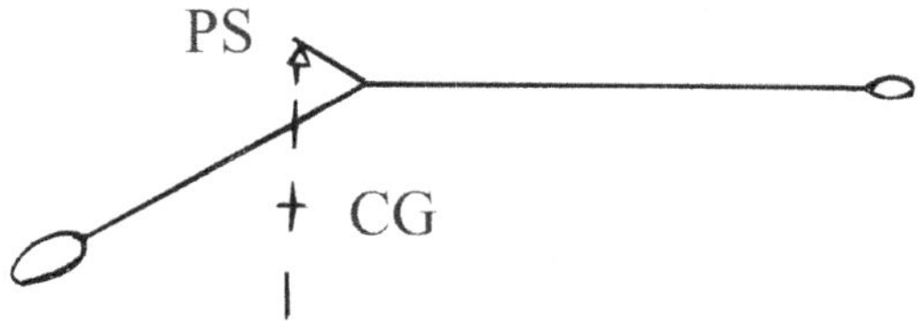

Fig. 3.51. A horizontal puppet in stable equilibrium with modeling clay in both hands and feet.

point of suspension PS (point of contact between our finger and the puppet), as in Fig. 3.50.

Any rotation of the puppet around the PS raises the CG. Gravity causes the CG to fall to the ground, with the puppet seated on our finger once more.

When we put clay on the hands and feet of the puppet, so that it stays lying down in a horizontal position, supported by our vertical pointing finger under its head, the CG is also located vertically below the point of suspension. In this case it is difficult to locate the CG exactly. But in Fig. 3.51 we show a deformed puppet in order to illustrate the location of the CG.

The body is horizontal, the head is raised a little, the arms are inclined downward a little, and the clay is placed on the hands and feet. The point of suspension PS is represented by a small triangle placed below the head. The CG is located vertically below the PS.

All the phenomena observed with the equilibrist can be explained with the basic experimental data and properties of the CG we have already presented. But it is extremely important that all students perform these experiments themselves in the classroom, each with his own equilibrist and clay, because this creates a deep impression upon them. The feelings of mystery and awe stimulated by this experiment are really

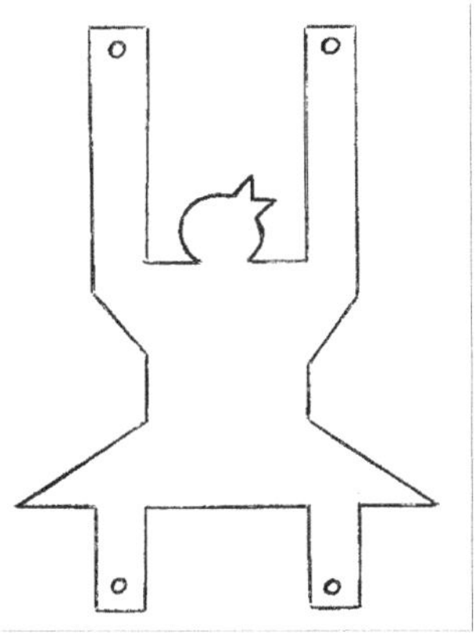

Fig. 3.52. A female equilibrist.

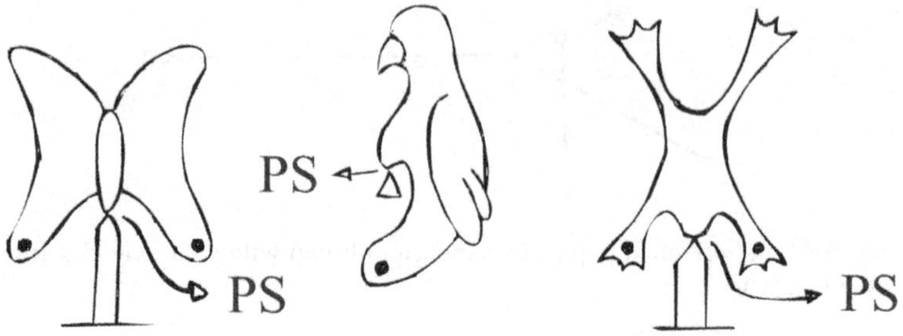

Fig. 3.53. A butterfly, a parrot and a frog.

remarkable. With this playful experiment they learn the main concepts relating to the CG.

3.4 Equilibrium toys

In addition to a male equilibrist, we can also make a female equilibrist, as in Fig. 3.52. Instead of using modeling clay on the hands and feet, we can also use fishing sinkers or other appropriate material. For a more durable figure, it is best to use thin sheets of wood or plastic instead of the pasteboard.

Other symmetric figures can also be made, such as a butterfly, a parrot or a frog. The black points in these Figures represent extra weights (modeling clay, for example), as in Fig. 3.53.

Some shops sell the bird equilibrist supported by its beak. Normally it is made of plastic, with shot hidden in the wings, and sometimes in the tail. It can also be made of pasteboard, as in Fig. 3.54.

In this case we put clay or small shot in the wings and tail until it remains in equilibrium horizontally, supported on a vertical stand under the beak. Most people believe that in this case the CG is on the end of the beak, where it touches the vertical stand. But as we have already

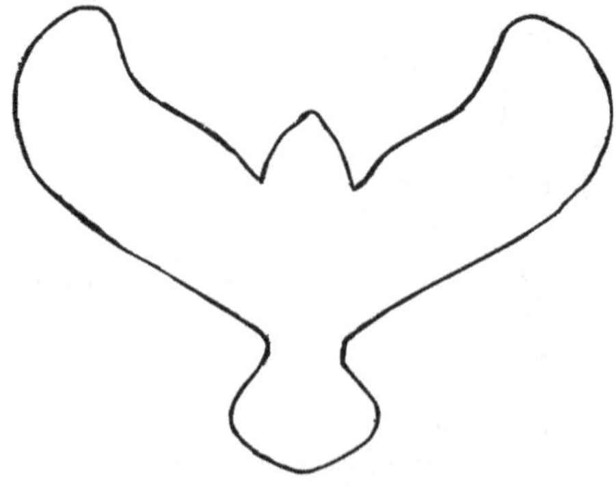

Fig. 3.54. A bird equilibrist.

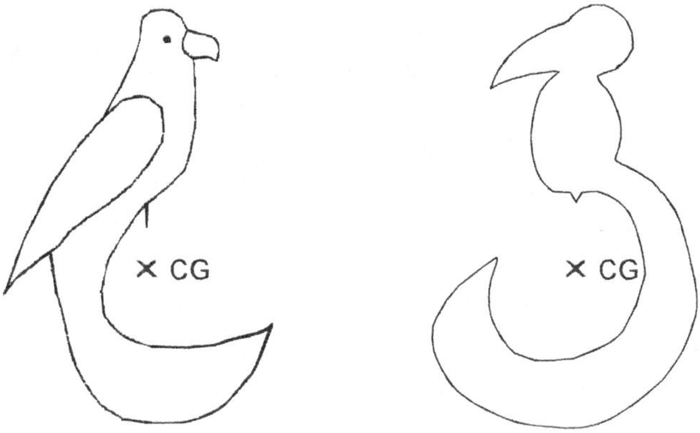

Fig. 3.55. The macaw and the toucan.

seen, in a situation of stable equilibrium the CG is not exactly at the beak, but a little below it, between the beak and the lower part of the shot in the wings. When we remove the bird from this equilibrium position (by raising or lowering one of its wings, or by lifting or bringing down its tail), releasing it from rest, it oscillates around the equilibrium position, its amplitude of oscillation decreasing due to friction, until it returns to the horizontal position. In this stable position the CG is in its lowest possible location.

The pasteboard equilibrist works exactly like this bird when the puppet is balanced horizontally with the pointing finger placed vertically under its head. The appropriate weights placed at the hands and feet of the puppet lower the CG of the system, so that in horizontal equilibrium it is vertically below the head. The advantage of the pasteboard equilibrist as compared with the bird bought in shops is that by changing the amount and location of the clay we can use the equilibrist both horizontally, like the bird, and sitting on our hands with the head at the top, or upside down balanced vertically on our finger.

There are also equilibrium figures made of homogeneous sheets of wood or plastic which do not require any additional weights. Some of

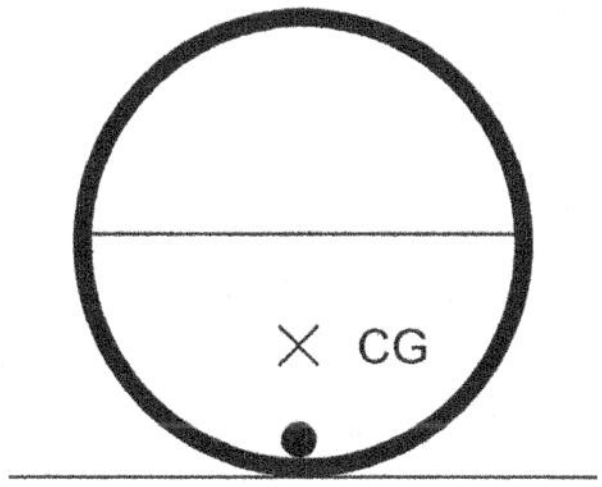

Fig. 3.56. The roly-poly doll.

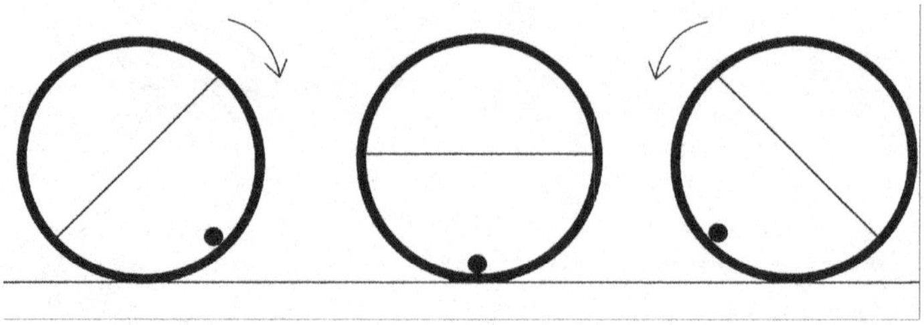

Fig. 3.57. Stable equilibrium of the roly-poly doll.

the most interesting examples are the macaw and toucan, as in Fig. 3.55.

These figures can also be made of rigid pasteboard. The foot can be a toothpick or a needle. In the toucan, the foot is only the pasteboard in the shape of a triangle. The important thing is that the macaw and toucan should have a large tail, such that the CG is located in the empty space between the foot and the tail. When this happens, the toucan remains balanced vertically supported by the tip of its foot. Any perturbation causes it to oscillate around this situation of stable equilibrium in which the CG is in its lowest position.

Another toy that is known to everyone is the roly-poly doll. It is based on the same principles that we have seen thus far. To build a roly-poly we need only two hemispheres or styrofoam spherical shells, plus some shot, modeling clay, or another weight. The CG of the homogeneous sphere is located at the center of the sphere. The CG of the extra weight is located at the center of the extra weight, assuming it is spherical in shape. When we place the shot at the bottom of one of the hemispheres, the CG of the whole system is located between the shot and the center of the sphere, as in Fig. 3.56.

This is the position of stable equilibrium for the roly-poly doll, as the CG for the whole system is at its lowest position. By tipping the roly-poly clockwise or anti-clockwise, we shift its CG away from the vertical, passing through the new point of support, raising the CG. Grav-

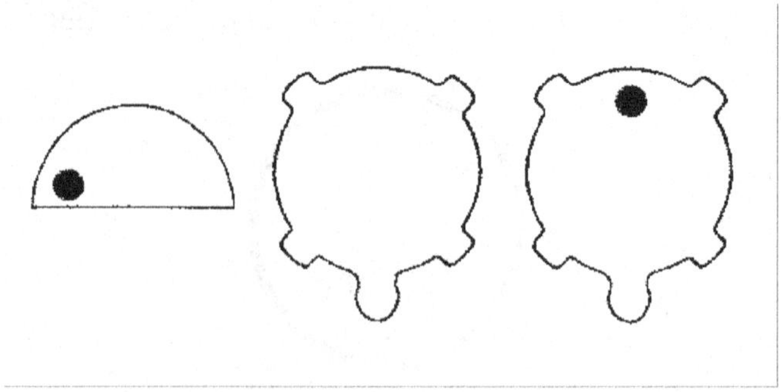

Fig. 3.58. The flip-flop turtle.

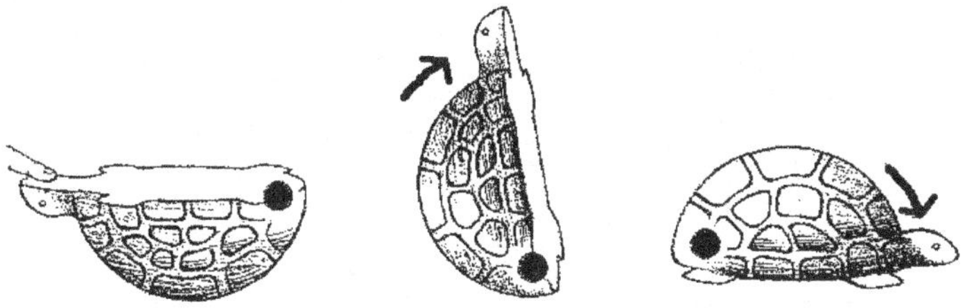

Fig. 3.59. The flip-flop turtle in action.

ity returns the doll to its stable position, as in Fig. 3.57.

The flip-flop turtle is another interesting toy, as in Fig. 3.58.[7] It is a different type of roly-poly in which the extra weight is placed asymmetrically relative to the equator of one hemisphere.

In this case we use only one hemisphere, the extra weight and a plane pasteboard figure having the same diameter as the hemisphere, but with four legs and a head simulating the shape of a turtle. The weight should be placed opposite to the head. We can hold the turtle upside down with its legs in a horizontal plane, pressing it by its chin. When we release it in this position, it turns over, returning to upright position, as in Fig. 3.59.

The reason for this is that the initial position is not a situation of equilibrium, because the CG is not in its lowest location. In the position of stable equilibrium the plane of the base and legs remains inclined relative to the vertical line. Small perturbations around this position cause the turtle to wobble around it. When we place it upside down with its base horizontal and release it from rest, it begins to move, lowering its CG. But as it acquires enough kinetic energy and we have only one hemisphere (unlike the roly-poly doll which has an external spherical or symmetrical shape relative to the position of equilibrium), the turtle turns over when the plane of the base and legs go beyond the vertical line.

3.4.1 Equilibrium games in the pub

Equilibrium games are often found in pubs and bars. All of them can be explained by the principles presented here. Even so the observed effects are very surprising and remarkable.

One of the most common is a needle or toothpick passing through the axis of a cork.[8] We then stick two metal forks in the cork, inclined

[7] Gaspar, *Experiências de Ciências* pp. 151-153.
[8] Gaspar, *Experiências de Ciências* pp. 144.

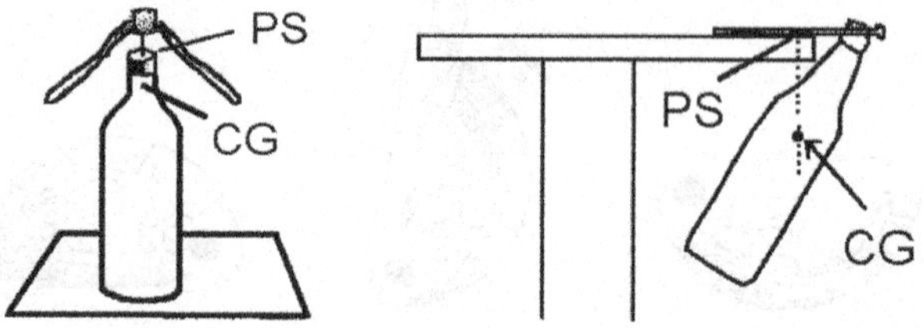

Fig. 3.60. Two interesting equilibrium situations.

downward toward the tip of the needle. The whole system can be balanced by placing the tip of the needle above a bottle, as in Fig. 3.60.

Most people think that the CG is at the tip of the needle. However, as a matter of fact, the tip of the needle is only the point of suspension PS of the system. In stable equilibrium, as we saw before, the CG is located vertically below the PS. In order to show that this is a condition of stable equilibrium, we can blow lightly on one of the forks so that the system turns around the vertical axis. It is also possible to blow vertically from top to bottom on one of the forks (or to lower it a little with our finger, releasing it from rest). The system will oscillate around the horizontal plane, finally stopping at the equilibrium configuration.

Another interesting situation is a full bottle, with cap, supported at the edge of a thin table by a bottle opener, as in Fig. 3.60.[9] The PS along the plane of the bottle opener is once again vertically above the CG along the axis of symmetry of the bottle. To try this experiment, it is wise to place a pillow or cushion below the bottle. This will prevent it from breaking if it falls to the ground while we are performing the experiment.

One of the most remarkable experiments utilizes a metal fork with its teeth connected to a spoon. A toothpick is passed partly through the teeth of the fork. At this point we can equilibrate the system by placing

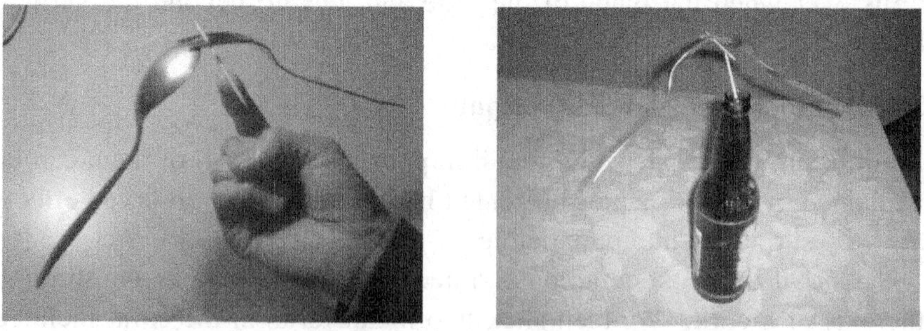

Fig. 3.61. Another curious equilibrium situation.

[9] Gaspar, *Experiências de Ciências* pp. 144.

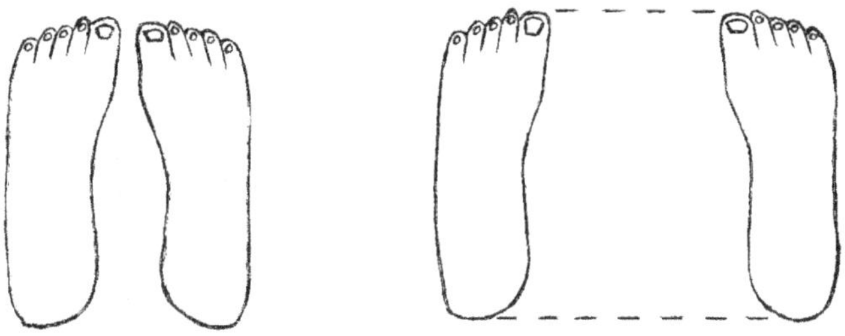

Fig. 3.62. Region of equilibrium for a standing person.

our pointing finger vertically under the toothpick, as in Fig. 3.61.

This localizes the appropriate PS for the toothpick. We can then go on with the game. We now support a second toothpick in the mouth of an open bottle. Supporting it firmly with our hands, we place the first toothpick with the PS above the upper tip of the second toothpick. With a little practice we can then finally release the system so that it remains in equilibrium, as shown in Fig. 3.61.

Once more the CG of the system is located vertically below the PS. The amazing fact about this game is that the PS is supported by a single point, namely, the upper tip of the second toothpick. Many people are surprised by this equilibrium, because they incorrectly believe that the CG is exactly at the point of contact of the two toothpicks. Moreover, this is a highly stable equilibrium. In order to show this, we only need to blow on the spoon horizontally, so that the system turns around the vertical direction passing through the PS. It is also possible to blow on the spoon from above (or lower it slightly with our finger, releasing it from rest). In this case the system oscillates around the horizontal plane, coming back to the original position of equilibrium.

3.4.2 Equilibrium of the human body

Several interesting experiments can be made on equilibrium of the human body. The legs and arms of a person can move independently from the rest of the body, moving forwards, backwards, upwards or downwards. All these movements change the location of the person's CG.

Let us consider initially the situation in which a person is standing above a flat surface. The CG is located above the ground. As we have seen before, equilibrium is only possible in this case when the CG is vertically above the region of support. When a person is standing, the CG is approximately at the center of the chest. The person can then remain in equilibrium in this situation provided the vertical projection of the CG falls inside the region bounded by the feet, as in Fig. 3.62.

Fig. 3.63. Equilibrium of the human body.

When the person stands with feet spread apart (wide support base), this region expands, as in Fig. 3.62.

In the first game we ask the person to bend at the waist, keeping knees straight, and touch the toes. Once this has been done, we ask the person to repeat the procedure, but this time, standing with the back up against a wall, with buttocks and heels touching the wall. This time it can't be done, as in Fig. 3.63. In order to understand what happens, it is best to ask the student to stand at the side of the classroom, in profile. The situation can be depicted on the blackboard. When the person is

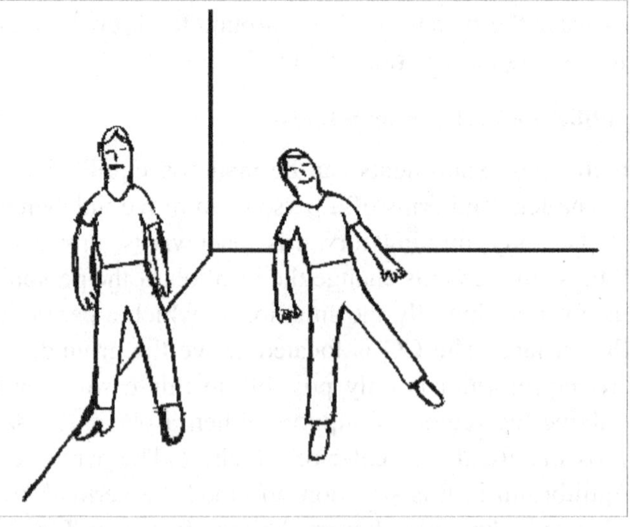

Fig. 3.64. Another situation of equilibrium of the human body.

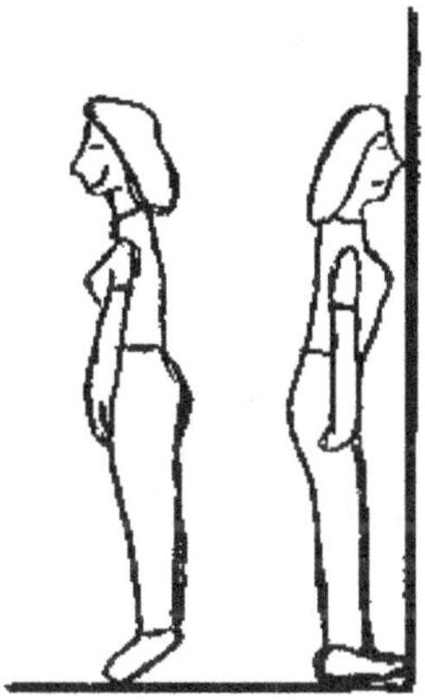

Fig. 3.65. Equilibrium by standing on the toes.

standing, the downward vertical projection of the CG passes through the feet. The person can only touch the toes by moving the buttocks backward and the head forward, in such a way that the projection of the CG continues to fall through the region enclosed by the feet. Now suppose the person stands with the back up against a wall. The person can no longer bend fully at the waist. When the arms and waist are lowered, the vertical projection of the CG falls outside the region between the feet, because the wall prevents the buttocks from moving backwards. The person looses equilibrium and falls forward.

Another game involves raising the left foot to the side while standing on the right foot. Everyone can do this. We then ask for the person to repeat the procedure, but now standing with the right shoulder and right foot against the wall. No one can raise the left foot and stand on the right foot for a few seconds in this new situation, as in Fig. 3.64. The explanation is the same as in the previous case. When the person is standing with both feet on the ground, the vertical projection of the CG falls between the feet. The person can only balance on the right foot while raising the left foot to the one side by leaning to the opposite side, in such a way that the vertical projection of the CG falls over the foot which is on the ground. Now let us consider the case in which the person is standing with the right shoulder and right foot against the wall. When the person lifts the left foot, the body has a tendency to move to the opposite side, in order to maintain balance. But the rigid wall prevents the upper part of the body from moving. The vertical projection of the CG when the left foot is raised to the side now falls outside the re-

Fig. 3.66. Equilibrium by kneeling down.

gion of the right foot. The CG then starts moving toward the ground, the person looses balance, and cannot complete the movement.

A third game is based on the same principle. We ask the person to stand on the toes. Everyone can do this. We then ask the person to repeat the procedure, but now standing facing a wall, keeping the nose and toes touching the wall. Now the person cannot remain in equilibrium while standing on the toes, as in Fig. 3.65. The explanation is the same as in the other cases, but now with movements of smaller magnitude. That is, the wall prevents the forward motion of the body. When the person stands on the toes, the vertical projection of the CG falls behind the toes. The person looses equilibrium and can no longer stand on the toes for a few seconds.

One of the most interesting experiments of this kind shows a distinction in the location of the CG for men and women of the same height. As women have larger hips than men, their CG is a little lower than the CG of men. We ask a woman to kneel down and touch the elbows with the knees, with the hands on the ground, as if praying. We place a match box on the ground, touching the tip of her fingers. We then ask the woman to place her hands at the back and to try and knock down the match box with her nose, and then to come back to her initial position without touching the ground with her hands, as in Fig. 3.66.

The majority of women can do this after a few trials. But men cannot normally do this. Let us consider the situation where the woman

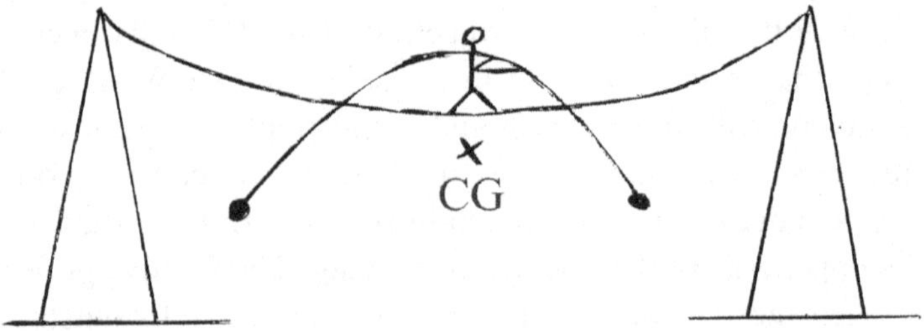

Fig. 3.67. Acrobat on a tightrope.

touches the match box with her nose. The vertical projection of her CG falls over the region occupied by her knees and feet. The CG of standing men is normally higher than the CG of standing women of the same height. If we suppose a man touching the match box with his nose, the vertical projection of his CG falls outside the region occupied by his knees and feet, and inside the region between the knees and the match box. As the tendency of the CG is to fall when there is no support below it, the man looses balance and cannot knock the match box down. If he tries to do this, he will fall to the ground and will not come back to his original position with his hands at his back, without first touching the ground with his hands.

Other situations of equilibrium occur when the CG is below the point of suspension PS. The most interesting example is an acrobat on a tightrope in a circus. The CG of a person is normally in the middle of the chest. If the person is standing above a tight rope, it is difficult to keep the projection of the CG from falling above the small region occupied by the feet. Normally this is done by a continuous deformation of the body in order to achieve balance. An alternative procedure is to hold a long curved stick with weights at the tips, as in Fig. 3.67.

The goal of this curved stick is to lower the CG of the system (person plus stick) below the feet. Any disturbance in the person's position will raise the CG. This will happen not only for clockwise and anticlockwise rotations, but also when the person leans forward or backward. As the tendency of the CG is to fall, the equilibrist ends in stable equilibrium, in which the acrobat stands vertically above the CG.

This is the ideal situation of equilibrium for rigid bodies, as in some toys. For a real acrobat in a circus, the stick is sometimes straight and the CG of the system may be located above the feet of the tumbler. The person tends to fall after any disturbance. In order to maintain balance, the acrobat needs to be constantly in motion, bending and stretching his body in order to keep changing all the time the position of his CG. When the person is falling to one side, he moves the stick to the other side. The person and stick need to stay constantly in motion.

The fun game we played with the pasteboard equilibrist presents a situation analogous to this for a rigid body. Normally we cannot keep the pasteboard in balance seated on our finger. But when we place enough clay on the feet of the equilibrist, we can keep it balanced on our finger, with the body of the equilibrist in a vertical plane. No matter which direction it wobbles, it always returns to the position of stable equilibrium. In this condition the CG is in the lowest position, vertically below the PS.

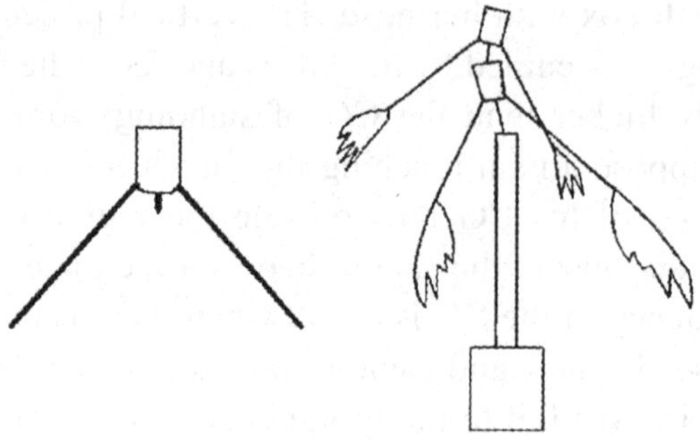

Fig. 3.68. The ET.

3.4.3 The ET

Another curious toy is the ET.[10] It can be made with two corks, two toothpicks, and four bamboo barbecue skewers, pieces of pasteboard for the hands and feet, plus a vertical stand to support it. Instead of the toothpicks we can also employ nails or needles.

The ET has two independent parts. If one of the corks is smaller than the other, it should be utilized in the upper part. We pass a toothpick, nail or needle through the axis of the cork. The bamboo barbecue skewers will form the arms of the ET, when inserted into the cork. They should be inclined downward, to the same side where the toothpick is pointing outward. This will also be the general shape of the body and legs of the ET, as in Fig. 3.68.

On the outer tips of the bamboo skewers we attach pieces of pasteboard in the shape of hands. After finishing the upper part of the ET, we try to balance it on our finger placed under the toothpick. If it falls to one side, we can increase the weight or size of the hands, or we can change the inclination of the bamboo skewers by placing them closer to the vertical direction (in order to lower the CG of the upper part). The important point is that the CG of the upper part should be below the lower tip of the toothpick in order to achieve stable equilibrium.

The lower part of the ET is made similarly. We may have to increase the weight of the feet relative to the weight of the hands in order to significantly lower the CG of the whole system. Once more, the lower part should be well balanced in a vertical plane before we proceed with the experiment.

We can then support the upper part of the ET on the lower part, by balancing the upper toothpick on the lower cork. We next support the

[10] Ferreira, *Equilíbrio*

lower toothpick on a rigid support. The final setup should be similar to Fig. 3.68.

This puppet is not a rigid body, as its two parts are free to wobble and turn independently of one another. Nevertheless each part of the ET can be considered, separately, a rigid body. By rocking or blowing the ET we produce some very curious motion.

Each one of these parts will only be balanced if its CG is below the upper tip of its toothpick. Moreover, the CG of the whole ET must be located below the bottom tip of the lower toothpick. Nevertheless, there are two possible alternatives. In the first, the CG of the upper part is below the bottom tip of the lower toothpick. In the second, the CG of the upper part is above the bottom tip of the lower toothpick.

This is an amusing toy that can raise many questions from the students.

In the next Chapter of this book we will see several definitions of the CG that have been presented through the centuries. We will see that it has always been difficult to find appropriate words to define the CG in a general way. Several important authors have dealt with this subject. In a latter Chapter of this book we will deal with the theoretical calculation of the CG. For the time being it is important to keep in mind the general definition CG8 and the practical procedures to locate the CG given by CG6 and CG7.

Chapter 4

Historical Aspects of the Center of Gravity

We now discuss a few historical aspects of the concept of center of gravity, CG. In particular, we will analyze how the concept was defined and how it was obtained experimentally. We are interested in the period in which this concept originated and was established. The information here is drawn essentially from the original works of Archimedes, Heron and Pappus, and from the books by Heath, Duhem and Dijksterhuis.[1]

The observation that bodies fall to the ground when released from rest above the Earth is extremely old. The same can be said of the fact that rigid bodies can remain in equilibrium after release when they are supported by a rigid stand placed below some specific point. It is probable that all ancient civilizations knew this. Nevertheless, the systematic and scientific treatment of the conditions which determine the equilibrium of bodies upon the surface of the Earth originated in Greece. At least Greece is the origin of the oldest documents dealing with the CG that give theoretical results on the subject.

Archimedes is the main person who investigated this concept in ancient Greece. The CG is also called barycenter. The prefix "bary" is a Greek root meaning weight or heavy. The literal meaning of the word barycenter is "center of weight." The simplest way to understand this expression and the concept behind it is to observe the experiment where we supported a pasteboard triangle in a horizontal plane, standing on a

[1] T.L. Heath, ed., *The Works of Archimedes* (New York: Dover, 2002); T.L. Heath, *A History of Greek Mathematics*, Vol. II: *From Aristarchus to Diophantus* (Oxford: Clarendon Press, 1921); C. Mugler, *Les Oeuvres d'Archimède* (Paris: Budé, 1970-72). Vol. 1-4; Pappus d'Alexandrie, *La Collection Mathématique* (Paris: Blanchard, 1982); Héron d'Alexandrie, *Les Mécaniques, ou, L'Élévateur des Corps Lourds* (Paris: Les Belles Lettres, 1988); E.J. Dijksterhuis, *Archimedes* (Princeton: Princeton University Press, 1987); P. Duhem, *The Origins of Statics* (Dordrecht: Kluwer, 1991).

vertical support placed under its centroid. The triangle only remains in equilibrium after released from rest when supported by this point. The whole weight of the figure is supported by this point, as if it were concentrated in it. It is then natural to call this specific point the center of weight, or barycenter, of the triangle.

The oldest extant work of Archimedes is called *On the Equilibrium of Planes*, or *On the Center of Gravity of Planes*.[2]. The center of gravity appears in postulates 4 to 7, without any prior definition:

> Postulate 4: When equal and similar plane figures coincide if applied to one another, their centres of gravity similarly coincide.
>
> Postulate 5: In figures which are unequal but similar the centres of gravity will be similarly situated. By points similarly situated in relation to similar figures I mean points such that, if straight lines be drawn from them to the equal angles, they make equal angles with the corresponding sides.
>
> Postulate 6: If magnitudes at certain distances be in equilibrium, (other) magnitudes equal to them will also be in equilibrium at the same distances.
>
> Postulate 7: In any figure whose perimeter is concave in (one and) the same direction the centre of gravity must be within the figure.[3]

In all likelihood the CG had been defined by Archimedes in one of his other works on mechanics that is no longer extant, namely, *On the Centers of Gravity*, *Elements of Mechanics*, *On Equilibria*, *On Balances* or *On Levers*, and *Book of Supports*.

In Proposition 6 of his work *Quadrature of the Parabola*, Archimedes writes:[4] "Every body, suspended by any point, assumes an equilibrium state when the point of suspension and the center of gravity are on the same vertical line. This has been demonstrated." This shows that Archimedes's knew the practical procedure CG6 of how to find the CG experimentally. That is, we suspend the rigid body by a point of suspension PS_1, wait until the body reaches equilibrium and draw the vertical passing through the PS_1 with the help of a plumb line. We suspend the body by another point of suspension PS_2 which is not along the first vertical, wait until it reaches equilibrium, and draw a second vertical through PS_2. The intersection of the two verticals is the CG of the body. But it is important to emphasize that to Archimedes this was not a definition of the CG. Instead, he proved this result theoretically utilizing

---

[2] Heath, ed., *The Works of Archimedes*, p. 189; Dijksterhuis, *Archimedes*, p. 286.

[3] Heath, ed., *The Works of Archimedes*, pp. 189-190.

[4] Mugler, *Les Oeuvres d'Archimède*, Vol. 2, p. 171; Duhem, *The Origins of Statics*, p. 463.

a previous definition of the CG of a body, as well as some postulates that are now unknown.

The crucial sentence in the previous paragraph does not appear in its full generality in Heath's translation of Archimedes's work. Heath's work is a paraphrase, that is, it conserves Archimedes original ideas, but rephrases them in modern notation and omits parts of the text which he did not consider essential. Here is Heath's presentation of Archimedes's key Propositions 6 and 7 of his work *Quadrature of the Parabola*.[5] In these Propositions the expression ΔBCD means the area of the triangle BCD, which is supposed to have uniform density. That is, its weight is proportional to its area, the same holding for the area P of the rectangle, which he will use in this Proposition.

> Propositions 6, 7[*]
>
> Suppose a lever AOB placed horizontally and supported at its middle point O. Let a triangle BCD in which the angle C is right or obtuse be suspended from B and O, so that C is attached to O and CD is in the same vertical line with O. Then, if P be such an area as, when suspended from A, will keep the system in equilibrium,
>
> $$P = \frac{1}{3}\Delta \mathrm{BCD}\,.$$
>
> Take a point E on OB such that BE = 2OE, and draw EFH parallel to OCD meeting BC, BD in F, H respectively. Let G be the middle point of FH.
>
> Then G is the centre of gravity of the triangle BCD.
>
> Hence, if the angular points B, C be set free and the triangle be suspended by attaching F to E, the triangle will hang in the same position as before, because EFG is a vertical straight line. "For this is proved[†]."
>
> Therefore, as before, there will be equilibrium.
>
> Thus
>
> $$P : \Delta \mathrm{BCD} = \mathrm{OE} : \mathrm{AO} = 1 : 3\,,$$
>
> or
>
> $$P = \frac{1}{3}\Delta \mathrm{BCD}\,.$$

[5] Heath, ed., *The Works of Archimedes*, p. 238.

[*] Note by Heath: In Prop. 6 Archimedes takes the separate case in which the angle BCD of the triangle is a right angle so that C coincides with O in the figure and F with E. He then proves, in Prop. 7, the same property for the triangle in which BCD is an obtuse angle, by treating the triangle as the difference between two right-angled triangles BOD, BOC and using the result of Prop. 6. I have combined the two propositions in one proof, for the sake of brevity. The same remark applies to the propositions following Props. 6, 7.

[†] Note by Heath: Doubtless in the lost book περί ζυγων. Cf. the Introduction, Chapter II., *ad fin*.

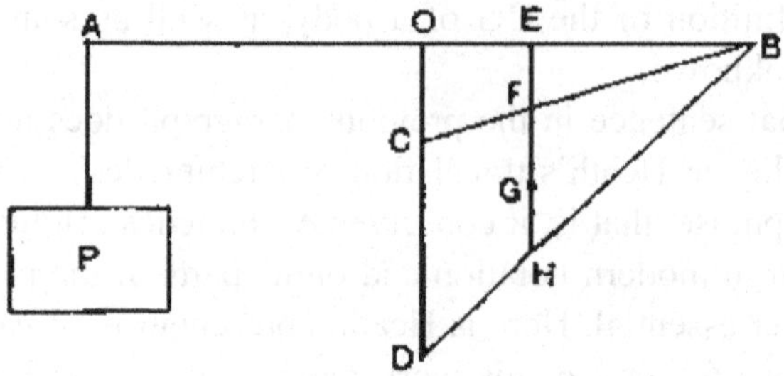

Eutocius of Ascalon (480-540) wrote commentaries on three works by Archimedes: *Measurement of a Circle*, *On the Sphere and Cylinder*, and *On the Equilibrium of Planes*. Apparently he did not know the other works. In his comments on book I of the equilibrium of planes, Eutocius clarifies a few points regarding the CG. These ideas are from Eutocius, not Archimedes, but are interesting nevertheless. We translate them from the French version published by Charles Mugler in 1972, which is a literal translation from the Greek:[6]

> Commentaries of Eutocius relative to Book I of Archimedes's work On the Equilibrium of Plane Figures. Introduction to book I. (...) In this work Archimedes defines the center of motion of a plane figure as the point such that, when we suspend the figure by this point, it remains parallel to the horizon, and defines the center of motion or of gravity of two or of several plane figures as the point such that, when we suspend the figures by this point, the beam (connecting the figures) remains parallel to the horizon.
>
> Let, for instance, ABΓ be the triangle and inside it the point Δ, such that when the triangle is suspended by this point, it remains parallel to the horizon. Therefore, it is clear that the parts B and Γ of the triangle balance one another and that none of them inclines more than the other relative to the horizon. In the same way, let AB be the beam of a balance and A and B the magnitudes suspended by it. If the beam, being suspended by Γ, keeps A and B in equilibrium, and remains parallel to the horizon, Γ will be the point of suspension of the magnitudes A and B.

These are clear and intuitive definitions, as we saw in the experiments presented earlier. But they are limited, because they do not deal with concave or pierced figures, for which the CG is located in empty space. Moreover, they do not apply to three-dimensional bodies. In spite of this, they illustrate many important aspects of the CG. It is also interesting to see the alternative expressions utilized for the CG: center of motion and point of suspension.

[6] Mugler, *Les Oeuvres d'Archimède*, Vol. 4, pp. 166-7.

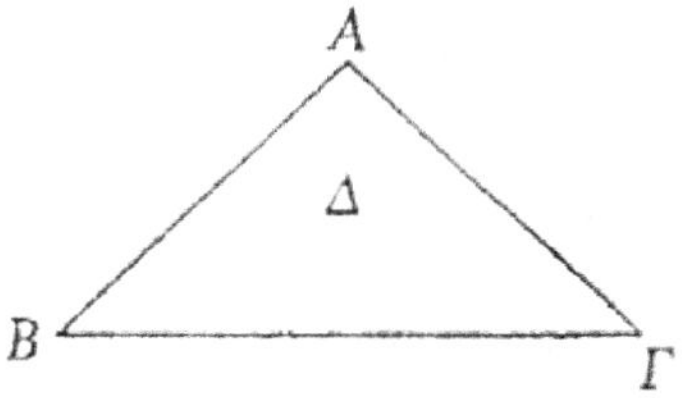

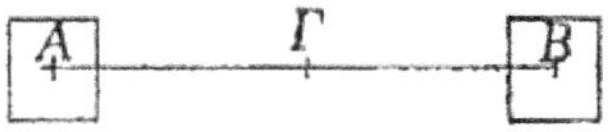

To form an idea how the concept of CG might had been defined by Archimedes, we quote here a few passages from the work *Mechanics* by the mathematician Heron (first century A.D.), from the *Mathematical Collection* by the mathematician Pappus (fourth century A.D.), and from the *Commentaries* of the philosopher Simplicius (sixth century A.D.), about *On the Heavens*, of Aristotle (384-322 B.C.). These authors discussed Archimedes's works, quote some of his works no longer extant and, probably, follow his concepts and lines of reasoning when dealing with the barycentric theory.

There is much controversy about the period in which Heron of Alexandria lived, but nowadays it is agreed that he flourished in the first century A.D. There are only fragments of his book *Mechanics*, in three parts, in Greek. But a complete Arabic translation of this work has been preserved. Translations have been made to other modern languages (French, in 1893, and German, in 1900) from this Arabic version.

Heron presents a definition of the CG as given by the stoic Poseidonius, who probably lived before Archimedes: "The center of gravity or of inclination is a point such that, when the weight is suspended by this point, it is divided in two equivalent portions."[7] Heath translates this sentence as: "The centre of gravity or of inclination is a point such that, if the body is hung up at it, the body is divided into two equal parts."[8]This definition is vague and problematic. In the first place it is difficult to know how a point, or even a vertical line passing through this point (if we interpret Poseidonius sentence thus) can divide a three-dimensional body into two parts. Even if the body is a plane figure, a point will not divide it into two parts. And a straight line will only divide a plane figure into two parts if it lies in the same plane as the figure. Therefore, we would need to imagine a triangle, for instance, suspended in a vertical plane. But even in this case not all verticals passing

[7] Héron d'Alexandrie, *Les Mécaniques, ou, L'Élévateur des Corps Lourds* (Paris: Les Belles Lettres, 1988), Chap. 24, p. 93.

[8] Heath, *A History of Greek Mathematics*, p. 350.

through the CG will divide the triangle into two parts with equal area or with equal weight. Let us suppose a homogeneous triangle suspended in a vertical plane. We have seen before that a straight line passing through the CG and through a vertex divides the triangle into two parts with the same area and the same weight. On the other hand, a straight line parallel to the base of the triangle and passing through the CG does not divide the triangle into two equal parts. Despite this fact, a triangle hanging in a vertical plane will remain in equilibrium after being released from rest if it hangs by the CG or by any other point which is vertically above the CG. The same problem will arise with Poseidonius's definition even if we interpret it as saying that the CG is a point such that, if the body is suspended from it, the body is divided by any vertical plane through the point of suspension into two equal parts. In this case we can imagine a triangle equilibrated in a horizontal plane, supported on a vertical plane placed below it (as a matter of fact the support needs to have a small thickness, like the edge of a ruler). If the vertical plane passes through a vertex and the CG, the body will remain in equilibrium and the upward projection of this plane will divide the triangle into two equal areas or into two equal weights. On the other hand, if the vertical plane is parallel to the base of the triangle and passes through the CG, its upward projection will not divide the triangle into two equal areas nor into two equal weights. Yet the triangle will also remain in equilibrium after being released from rest.

Another expression utilized by Heron to designate the CG, apart from "center of gravity," is "center of inclination" or "center of fall." This expression was probably also utilized in ancient Greece. This is a very interesting and instructive expression. We saw that any body heavier than air tends to fall toward the ground when released from rest. If the body is suspended by a PS and released from rest, so that it can turn around this point, the initial motion of the CG (supposing that it does not coincide with the PS) is toward the ground. Therefore, it behaves as if the tendency to fall were concentrated at the CG of the body.

Heron then says that Archimedes distinguished between the "center of inclination" and the "point of suspension." Heron continues: "*The point of suspension is a point on the body or on a non corporeal figure such that, if the body is hung up at this point, its parts remain in equilibrium, that is, the body does not oscillate or incline in any direction.*"[9] The expression "non corporeal figure" may mean the cases in which the CG is located in empty space, like the CG of a ring. What is called the

---

[9] Héron d'Alexandrie, *Les Mécaniques,* Chap. 24, p. 93; Heath, *A History of Greek Mathematics,* p. 350.

"point of suspension" here and the definition Heron gave of it may have been how Archimedes defined the center of gravity. Later on we will see an analogous definition by Pappus.

Heron also writes: "The center of inclination in each body is one single point to which converge all the vertical lines through the points of suspension. The center of gravity in certain bodies is outside the substance of these bodies; this is what happens, for instance, in arches and in bracelets. All the lines following the projections of the ropes converge at a common point."[10] Here he seems to be describing the practical procedure CG6, *i.e.*, to find the CG through the intersection of all verticals passing through the points of suspension when the body is in equilibrium, at rest relative to the Earth. This is the most important practical way to locate the CG. Heron mentions that this CG may be located in empty space, outside the substance of the bodies, as is the case for rings or wheels.

Heron mentions that Archimedes solved problems like the following in his book *On Columns* or *On Supports*:[11] A heavy beam or a wall supported on a number of pillars, equidistant or not, even or not even in number, and projecting or not projecting beyond one or both of the extreme pillars, finding how much of the weight is supported on each pillar. Heron also says that the same principles can be applied when the body (beam or wall) is suspended by cables. In another part of his book Heron considers the problem of a triangle of uniform thickness, with its plane horizontal, supported by a pillar under each vertex. He then finds the weight supported by each pillar in several cases: (a) when they support only the triangle, (b) when they support the triangle plus a given weight placed at any location over the triangle. Then Heron finds the CG of the system when known weights are placed over the vertices of the triangle. He then extends his analysis to polygons.

Pappus presents an explicit definition of the CG, namely: "*We say that the center of gravity of any body is a point within that body which is such that, if the body be conceived to be suspended from that point, the weight carried thereby remains at rest and preserves its original position.*"[12] In another context he writes: "It is also clear that, if we imagine the body suspended by its center of gravity, it will not turn and will remain at rest preserving the initial position it acquired by the solicita-

[10] Héron d'Alexandrie, *Les Mécaniques*, Chap. 24, p. 95.

[11] Héron d'Alexandrie, *Les Mécaniques*, Chaps. 25-31; Heath, *A History of Greek Mathematics*, p. 350.

[12] Pappus d'Alexandrie, *La Collection Mathématique*, Book 8, p. 815; Dijksterhuis, *Archimedes*, p. 299.

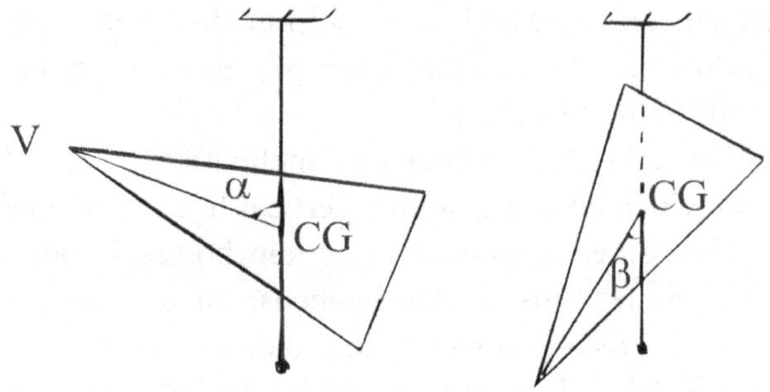

A vertical triangle suspended through its CG remains in equilibrium whatever the angle α. The same happens for any angle β when the plane of the triangle is inclined relative to the ground in an arbitrary manner, provided it is suspended by the CG.

tion."[13] By solicitation here he means tendency to fall toward the Earth due to gravity.

Simplicius quotes a similar definition by Archimedes: "*The centre of gravity is a certain point in the body such that, if the body is hung up by a string attached to that point, it will remain in its position without inclining in any direction*." (quoted by Heath)[14]

We can illustrate this definition with a triangle suspended by its CG. This is a thought experiment, as it may be impossible to suspend a body exactly by its CG, allowing the body to simultaneously rotate around this point in any direction. In any event, the idea is that if we could perform an experiment like this, what would happen is that the body would remain in equilibrium for all orientations in which it was released from rest. In the first situation we have a triangle in a vertical plane. The segment connecting one of its vertices and the CG makes an angle α with the vertical through the CG. If the triangle is released from rest, it will remain in equilibrium no matter the value of α. In the second situation the plane of the triangle (or a straight line orthogonal to it) is inclined relative to the vertical line. Let us call β the angle between the previous segment and the vertical through the CG. If the triangle is released from rest suspended by its CG, it will remain in equilibrium no matter what value β takes.

In a real experiment in which the body is suspended by a point of suspension PS which does not coincide with the CG, being free to rotate around the PS, the body will only remain in equilibrium after being released from rest if it is in the preferential position in which the PS and the CG are along a vertical line. If this is not the case, the body will turn

[13] Pappus d'Alexandrie, *La Collection Mathématique*, Book 8, p. 818.
[14] Heath, *A History of Greek Mathematics,* p. 350.

around the PS after release in such a way that the CG approaches the ground. In equilibrium the PS and the CG will be along a vertical line, with the CG below the PS.

Pappus describes a practical procedure for locating the CG.[15] He imagines a rectangular vertical plane over which a body will be suspended, balanced on the upper horizontal edge of the plane. The plane extended upwards divides the body into two parts which equilibrate one another. Next the body is supported over the same upper horizontal edge of the plane again, but this time with the body in a different orientation relative to the ground. The plane extended upward again divides the body into two parts which equilibrate one another. These two planes extended upward meet at a single vertical line. The body also remains in equilibrium when supported by a vertical line extended upward, as if it were supported at a point by a vertical stick. He repeats the procedure of balancing the body over the vertical rectangle in two new orientations of the body and obtains another vertical line (the intersection of these two new planes extended upwards). The intersection of both vertical lines is the CG of the body. According to Pappus, this is the most essential part of the barycentric theory. Moreover, Pappus says that the points demonstrated by this experiment were given in Archimedes's book *On Equilibria* and in Heron's book *Mechanics*. The procedure described by Pappus is analogous to our practical definition CG7. In other words, if a rigid body is supported at a point PA_1 by a vertical stick, the line of the stick extended upward (that is, the vertical V_1 passing through PA_1) must pass through the CG. We now imagine that the body with a new orientation relative to the ground is balanced at another point PA_2 by the same stick. The line of the stick extended upward is another vertical V_2 passing through PA_2. The intersection of these two verticals is the CG. This is analogous to the intersection of two verticals extended downward by two points of suspension, described by the practical procedure CG6.

Everything we have seen so far suggests that Heron, Pappus and Simplicius directly consulted certain treatises by Archimedes that are no longer extant. The underlined definitions by Heron, Pappus and Simplicius are analogous to our definition CG8. These authors also proposed practical procedures for locating the CG analogous to our CG6 and CG7.

[15] Pappus d'Alexandrie, *La Collection Mathématique.*, Book 8, pp. 816-818.

Chapter 5

Theoretical Values of Center of Gravity Obtained by Archimedes

Here we cite the theoretical values obtained by Archimedes for the centers of gravity of several one-, two- and three-dimensional figures. There are proofs of most of these results in the known works of Archimedes. In some cases, such as the CG of the cone, Archimedes gives only the results, stating that they had been proved previously. However, the calculations are not to be found in any of his extant works. It is presumed that he calculated them in another work which has been lost during the last two thousand years.

5.1 One-dimensional figures

A) "The centre of gravity of any straight line is the point of bisection of the straight line." (*The Method*)[1] In Heath this is Lemma 3, while in Mugler it is Lemma 4.

5.2 Two-dimensional figures

B) "The centre of gravity of a parallelogram is the point of intersection of its diagonals." (*On the Equilibrium of Planes*, Book I, Prop. 10)[2] "The centre of gravity of any parallelogram is the point in which the diagonals meet." (*The Method*)[3] In Heath this is Lemma 5, while in Mugler it is Lemma 6.

[1] Heath, ed., *The Works of Archimedes*, Supplement, p. 14; Mugler, *Les Oeuvres d'Archimède*, Vol. 3, p. 85.

[2] Heath, ed., *The Works of Archimedes*, p. 195.

[3] Heath, ed., *The Works of Archimedes*, Supplement, p. 14; Mugler, *Les Oeuvres d'Archimède*, Vol. 3, p. 85.

C) "The centre of gravity of any triangle is at the intersection of the lines drawn from any two angles to the middle points of the opposite sides respectively." (*On the Equilibrium of Planes*, Book I, Prop. 14)[4] "The centre of gravity of any triangle is the point in which the straight lines drawn from the angular points of the triangle to the middle points of the (opposite) sides cut one another." (*The Method*)[5] In Heath this is Lemma 4, while in Mugler it is Lemma 5.

D) "In any trapezium having two parallel sides the centre of gravity lies on the straight line joining the middle points of the parallel sides, in such a way that the segment of it having the middle point of the smaller of the parallel sides for extremity is to the remaining segment as the sum of double the greater plus the smaller is to the sum of double the smaller plus the greater of the parallel sides." (*On the Equilibrium of Planes*, Book I, Prop. 15)[6] Heath presents this proposition as: "If AD, BC be the two parallel sides of a trapezium ABCD, AD being the smaller, and if AD, BC be bisected at E, F respectively, then the centre of gravity of the trapezium is at a point G on EF such that $GE : GF = (2BC + AD) : (2AD + BC)$."[7]

E) "The centre of gravity of a circle is the point which is also the centre [of the circle]." (*The Method*)[8] In Heath this is Lemma 6, while in Mugler it is Lemma 7.

F) In Proposition 12 of *The Method* Archimedes finds the center of gravity of half a cylinder, that is, of a cylinder divided into two equal parts by a plane passing through the center of the cylinder. This result is analogous to obtaining the CG of a semicircle. See the discussion by Heath.[9]

G) "The centre of gravity of any segment comprehended by a straight line and an orthotome [parabola] divides the diameter of the segment in such a way that the part towards the vertex of the segment is half as large again as the part towards the base." (*On the Equilibrium of Planes*, Book II, Prop. 8)[10] Heath states this Proposition as follows: "If AO be the diameter of a parabolic segment, and G its centre of gravity, then $AG = (3/2)GO$."[11] Here A is the vertex of the parabolic segment.

---

[4] Heath, ed., *The Works of Archimedes*, Supplement, p. 201.

[5] Heath, ed., *The Works of Archimedes*, Supplement, p. 14; Mugler, *Les Oeuvres d'Archimède*, Vol. 3, p. 85.

[6] Dijksterhuis, *Archimedes*, p. 312.

[7] Heath, ed., *The Works of Archimedes*, p. 201.

[8] Heath, ed., *The Works of Archimedes*, Supplement, p. 15; Mugler, *Les Oeuvres d'Archimède*, Vol. 3, p. 85.

[9] Heath, ed., *The Works of Archimedes*, Supplement, pp. 38-40.

[10] Dijksterhuis, *Archimedes*, p. 353.

[11] Heath, ed., *The Works of Archimedes*, p. 214.

5.3 Three-dimensional figures

H) "The centre of gravity of any cylinder is the point of bisection of the axis." (*The Method*)[12] In Heath this is Lemma 7, while in Mugler it is Lemma 8.

I) "In any prism the center of gravity is the point of bisection of the axis." (*The Method*)[13] In Mugler this is Lemma 9. This lemma does not appear in Heath.[14] The "axis" here refers to the line segment joining the centers of gravity of the two bases, as appears from the application of this lemma in Proposition 13 of *The Method.*[15] A prism is a solid figure with similar, equal and parallel ends, and with sides which are parallelograms.

J) "The centre of gravity of any cone is [the point which divides its axis so that] the portion [adjacent to the vertex is] triple [of the portion adjacent to the base]." (*The Method*)[16] In Heath this is Lemma 8, while in Mugler it is Lemma 10.

K) "Let the axis of the segment of the paraboloid [of revolution] be AN (...) Let C be the centre of gravity of the paraboloid BAB′ (...) Then, since AN = ($\frac{3}{2}$)AC (...)" (*On Floating Bodies*, Book II, Prop. 2)[17] "The centre of gravity of a segment of a right-angled conoid (*i.e.*, a paraboloid of revolution) cut by a plane at right angles to the axis is on the straight line which is the axis of the segment, and divides the said straight line in such a way that the portion of it adjacent to the vertex is double of the remaining portion." (*The Method*, Prop. 5)[18] This is also discussed by Dijksterhuis.[19] That is, if the paraboloid of revolution has an axis of symmetry AN, with A being the vertex, the center of gravity C is located along AN in such a way that AC = 2CN, or AN/AC = $\frac{3}{2}$.

L) "The centre of gravity of any hemisphere [is on the straight line which] is its axis, and divides the said straight line in such a way that the portion of it adjacent to the surface of the hemisphere has to the remaining portion the ratio which 5 has to 3." (*The Method*, Prop. 6)[20]. That is, if the hemisphere has a radius R and its plane face is along the xy plane,

[12] Heath, ed., *The Works of Archimedes*, Supplement, p. 15; Mugler, *Les Oeuvres d'Archimède*, Vol. 3, p. 85.

[13] Mugler, *Les Oeuvres d'Archimède*, Vol. 3, p. 85.

[14] Heath, ed., *The Works of Archimedes*, Supplement.

[15] Dijksterhuis, *Archimedes*, p. 316, note 1.

[16] Heath, ed., *The Works of Archimedes*, Supplement, p. 15; Mugler, *Les Oeuvres d'Archimède*, Vol. 3, p. 85.

[17] Heath, ed., *The Works of Archimedes*, Supplement, pp. 264-5.

[18] Heath, ed., *The Works of Archimedes*, Supplement, p. 25.

[19] Dijksterhuis, 1987, p. 326.

[20] Heath, ed., *The Works of Archimedes*, Supplement, p. 27.

centered at the origin, the center of gravity will be along the z-axis (axis of symmetry) at a point z_{CG} such that $z_{CG} = 3R/8$.

M) "The centre of gravity of any segment of a sphere is on the straight line which is the axis of the segment, and divides this straight line in such a way that the part of it adjacent to the vertex of the segment has to the remaining part the ratio which the sum of the axis of the segment and four times the axis of the complementary segment has to the sum of the axis of the segment and double the axis of the complementary segment." (*The Method*, Prop. 9)[21]

N) Archimedes finds in Proposition 10 of *The Method* the center of gravity of any segment of an ellipsoid.

O) Archimedes finds in Proposition 11 of *The Method* the center of gravity of any segment of a hyperboloid of revolution.

The only point to be emphasized here is that all of these results were derived theoretically by Archimedes, beginning from his postulates. In other words, they were derived mathematically. In an earlier section of this book we saw how to obtain some of these results (such as the CG of a circle, rectangle, and triangle) experimentally. At the end of the book we will discuss how Archimedes calculated the CG of the triangle, as well as a modern mathematical definition of the CG.

[21] Heath, ed., *The Works of Archimedes*, Supplement, p. 35.

Chapter 6

Balances and the Measurement of Weight

By now we have arrived at a definition of the CG given by CG8, and two practical ways of finding it experimentally, CG6 and CG7. But these formulations do not enable us to calculate theoretically the CG of any discrete nor continuous distributions of matter. In order to do this we will need the concept of weight, a procedure to measure it, and also the law of the lever. This is our goal here.

We have seen in the experiments with the triangle in equilibrium, and in the geometrical analysis following it, that not all straight lines passing through the CG of a plane homogenous figure divide it into two equal areas. In the experiments with the pasteboard equilibrist we saw that by changing the location of the modeling clay attached to the equilibrist we could change the position of the CG of the whole system (pasteboard plus clay). This suggests that the CG has to do not only with the weight of the body, but also with the distribution of the matter around the body.

We will arrive here at a mathematical expression with which we can calculate the CG of any distribution of matter. To this end we need first to quantify the intuitive concept of weight. That is, to find a clear and objective way of measuring the weight of a body. This is the subject of the next few Sections.

6.1 Building a balance

The more basic or fundamental quantitative concepts we have in physics are those of the size of a body (or the distance between bodies), time between physical events, and weight of a body.

In order to measure the size of a body, or the distance between two bodies, we utilize essentially a rigid standard of length. By definition we say that two bodies have the same size when their extremities coincide. For example, we say that person A has the same height as person B if, when they are placed back to back, the heels and heads coincide with one another. By definition we say that body A is N times the size of body B if it is possible to superimpose in linear sequence N times body A between the extremities of body B. The simplest example of this is a 1 meter ruler divided into centimeters. We see that the ruler has 100 units of 1 cm between its ends, with these units stamped along the ruler. Utilizing a graduated ruler we can also measure the length of a body, or the distance between small bodies.

Time is the concept created by man in order to measure the changes which happen in nature. Any standard that repeats itself periodically can be utilized as a measure of time. Historically the most important and precise clock utilized in astronomy was the rotation of the Earth in relation to the background of stars seen with the naked eye. These stars are usually known as fixed stars, because they do not change noticeably their relative positions between one another while the Earth rotates relative to them. This leads to the definition of the unit of a sidereal day. Other astronomical clocks are given by the rotation of the Earth relative to the Sun, yielding the unit of a solar day, the phases of the Moon, or the variation of the position of the sunset in relation to the mountains and other terrestrial bodies, yielding the unit of a solar year. There are clocks with different degrees of precision. The simplest distinguish between darkness and light; others, the phases of the Moon, or the shadows of a gnomon. A gnomon is a vertical stick fixed in the ground which measures the height of the Sun in the sky through the orientation and size of its shadow. It is the basis of the construction of the sundials. Several other periodic phenomena have been utilized through the centuries to measure time: water clocks, mechanical clocks (based on a pendulum or a spring), electromagnetic clocks, atomic clocks, *etc.*

But the main concept we want to analyze in more detail here is the weight of a body. We all have an intuitive notion of the weight of a body as a quantitative measure of the gravitational force. We say that body A is heavier than body B if it is more difficult to keep body A in our hands at a certain height from the surface of the Earth than to keep body B at the same height. This difficulty can be indicated by our sweat, or by the fatigue we feel in our outstretched arm. We also say that body A is heavier than body B when we need to make a larger physical effort to raise body A to a certain height h than to raise body B to the same height h.

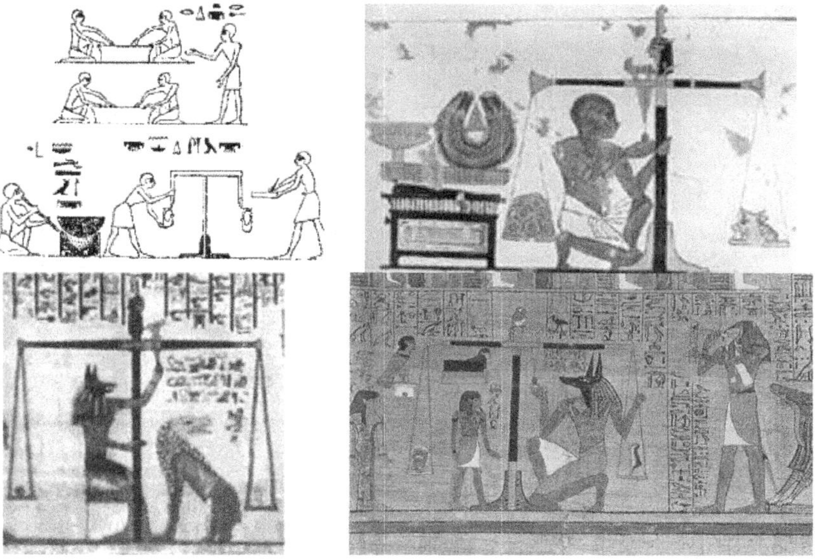

Fig. 6.1. Balances in ancient Egypt.

This sensorial and subjective notion can also be indicated by certain phenomena affecting other material bodies. For instance, the deformation caused by body A upon a material support holding it at rest relative to the ground. Let us suppose that this support is a spring. We can say that body A is heavier than body B if the same spring is more compressed supporting A than supporting B. In this case we would utilize a flexible and deformable body such as a spring as a weight indicator. It is better to use an objective phenomenon like the deformation of a spring in order to quantify the notion of weight than to use a subjective phenomenon like our sensation of fatigue.

But historically the oldest and most important instrument utilized to quantify the notion of weight was the equal arm balance. Balance is the name given to any instrument which determines quantitatively the weight of a body. The equal arm balance has been known since ancient Egypt, if not longer. In Fig. 6.1 are paintings from the time of the Pharaohs showing balances in use around 1500 B.C. It is interesting that three paintings show people holding a plumb bob to measure when the beam of the balance is horizontal.

According to Steve Hutcheon (private communication, which he obtained from Thompson[1]), the earliest record of the balance in Astronomy is from circa 1350 B.C. when the Akkadians of Mesopotamia

[1] G.D. Thomson, *The origin of the Zodiac*, available in 2007 at: http://members.optusnet.com.au/~gtosiris/page9a.html.

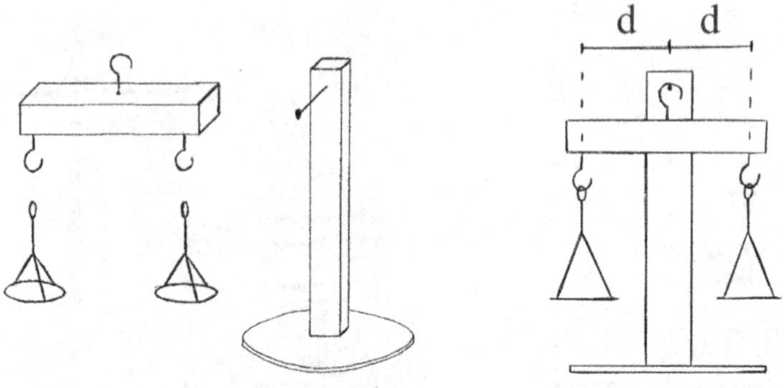

Fig. 6.2. Components of a balance.

called a star group Zibanitum (the scales). These stars later became known as the Zodiac constellation Libra. In that period Zibanitum gave the location of the Sun at sunrise on the Autumnal Equinox when the lengths of day and night, and the seasons, were in balance.

The main components of a balance are: (A) a homogeneous rigid beam free to rotate around a horizontal axis which is orthogonal to the beam, located halfway between the extremities of the beam (this axis is sometimes called the fulcrum of the balance), (B) a rigid support which keeps the fulcrum of the balance at rest relative to the ground, and (C) two scale pans, suspended at equal distances to the vertical plane passing through the fulcrum. An example is given in Fig. 6.2. The objects to be weighed are placed in these pans. The fulcrum may be part of the support, such as a horizontal needle fixed to the support, with the beam hanging from the needle. Or the fulcrum may be part of the beam, such as a horizontal needle fixed to the beam, with the needle supported by the fixed stand. We call the arm of the balance, the horizontal distance, d, between the point of suspension of the pan from the beam and the vertical plane passing through the fulcrum of the balance. In some balances we will build, no scale pans will be employed, as the bodies to be weighed will be suspended directly from the beam of the balance.

Before weighing any body, the balance must be adjusted in such a way that its beam is horizontal without the scale pans. If necessary, this can be done by changing the location of the fulcrum on the beam, or the length of the arm on one side of the beam. In addition, the beam must remain horizontal with the scale pans added. If necessary, this can be adjusted by changing the exact location of the nails where the scale pans are hanging. If the beam does not remain horizontal after these adjustments, we can sometimes succeed by placing a small counterweight at some point on one side of the beam. This counterweight can be a piece of thread, wire, or clay.

We have seen that by definition the direction followed by a falling body is called vertical, which coincides with the direction of a plumb line in equilibrium. Horizontal is any straight line or plane perpendicular to the vertical. We also defined the equilibrium of a body as the situation in which the body and all its parts remain at rest relative to the Earth.

Definition of a balance in equilibrium:

We define the meaning of the expression "balance in equilibrium" as the situation in which its arms remain at rest horizontally. This is the meaning given by most people to the equilibrium of balances, and we adopt it here. That is, even when the beam and scale pans are at rest relative to the ground, we will not say that the balance is in equilibrium if the beam is not horizontal.

Before utilizing the balance to measure weights, we must build it and to place it in equilibrium without the bodies to be weighed, only with its arms and scale pans. It is also important to verify that the threads holding the pans to the arms of the balance are at the same horizontal distance to the vertical plane passing through the fulcrum. In order to have a sensitive and precise balance, it is essential for it to be free to rotate around the fulcrum, without the hindrance of friction.

We have seen before that a rigid body suspended by a point is in stable equilibrium when the point of suspension PS is vertically above the CG of the body. If the CG is above the point of support PA, the equilibrium tends to be unstable, unless the PA is not a point but an area of support. For the time being we will deal only with balances suspended by a fulcrum located vertically above the CG of the empty balance. In some figures we will represent this fulcrum by the letters PS. One of the most important points in the construction of a balance is that the fulcrum must be vertically above the CG of the beam (without the scale pans and weights to be measured). This will guarantee the stable equilibrium of the beam. That is, it will return to the horizontal position after release from rest, no matter what the initial inclination of the beam relative to the horizontal.

The beam of the balance can be made of any rigid material like wood, plastic, metal, or even pasteboard. It can be cylindrical (a bamboo barbecue skewer or a broomstick), rectangular (a ruler or a rectangular pasteboard), or like a parallelepiped (a lath of wood). Close to the extremities of the beam, at equal distances to the vertical plane passing through the fulcrum, we can fix two equal nails, two needles or two

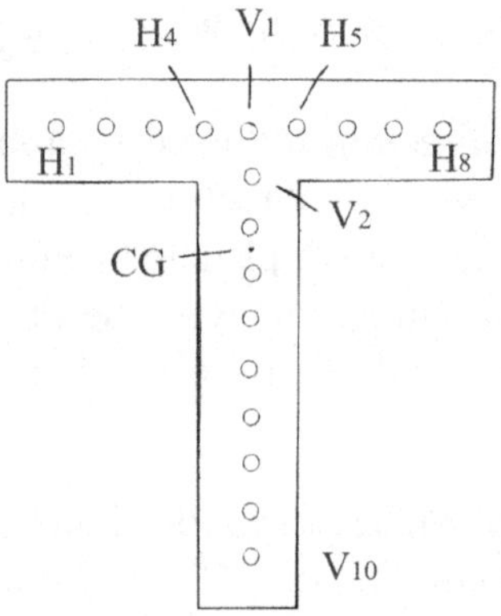

Fig. 6.3. A pasteboard balance.

hooks, which will support the scale pans by the threads. Instead of this, we can also make two equal holes at the extremities of the beam, where we will hang the pans (utilizing hooks). The pans of the balance may be two small bottle lids, two small plastic cups, or any other adequate supports for the bodies. We should make three holes symmetrically located around the edges of the lids or cups, where the threads will be tied. The threads holding the pans on each side of the beam should have the same length and should be made of the same material. Instead of lids or cups, we could also use small plastic or cloth bags suspended from the beam. Inside the bags we place the bodies to be weighted.

Here we present several types of equal arm balances that are very sensitive and precise, even though they are made with cheap and easy to find materials. They illustrate different possibilities, and can also be adapted to levers.

There are many ways in which the balance can have freedom of rotation around the fulcrum. One possibility is to have a hook on the upper part of the beam, above its midpoint. Another possibility is to make a horizontal hole halfway between its extremities, with the hole above the CG of the beam. In these two cases the balance hangs by a nail, needle, or bamboo skewer fixed horizontally in the rigid support, passing through the hook or hole made in the beam, as in Fig. 6.2.

One of the simplest balances found in every home is the coat-hanger. The cylindrical horizontal bar holding the coat-hanger is the fulcrum of the balance, and we can hang objects to be measured on the beam of the coat-hanger:

A simple and instructive balance model which we will utilize for some activities is made with a pasteboard in the form of the letter T. It has several holes along the body, as well as holes symmetrically located along the arms of the T, as in Fig. 6.3. By the method of the plumb line

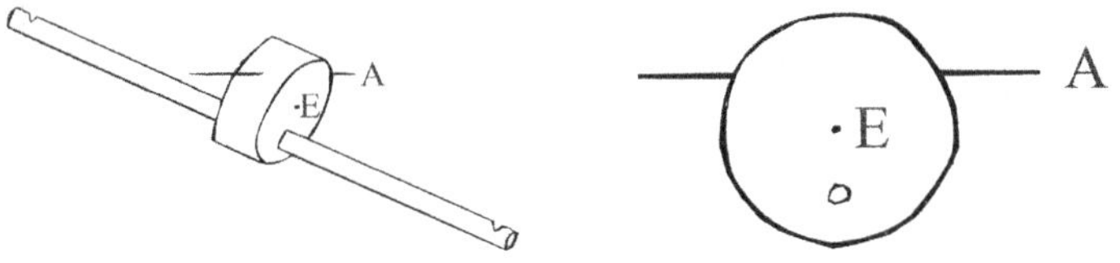

Fig. 6.4. The beam of a balance made with a bamboo barbecue skewer, a cork and a needle.

we can easily find the CG of the pierced figure. The T should be suspended by a fulcrum passing through a hole located above its CG. The scale pans can be suspended by any two holes along the arms, provided they are at equal distances from the vertical plane passing through the axis of symmetry of the T.

Another way to reduce friction from the wobbling of the balance is to use a horizontal stick fixed to the beam, orthogonal to it, supported on both sides by rigid, smooth stands of the same height. One example is made of a short cork (that is, a cork cut in half on a plane parallel to its faces), a needle (or toothpick), and a bamboo barbecue skewer. Initially we pass a nail longitudinally through the cork, parallel to its axis of symmetry, but outside it. We represent the axis of symmetry of the cork by the letter E. We remove the nail and pass the bamboo barbecue skewer through this hole. We then remove the tip of the bamboo skewer so that it becomes symmetrical. We then pass a needle or toothpick through the cork. The needle should be perpendicular to the bamboo skewer and parallel to the faces of the cork. We represent the needle by the letter A. The center of the cork must be located between the center of the bamboo skewer and the center of the needle. The bamboo skewer and the axis of the cork should be parallel, with the needle orthogonal to the plane formed by the axis and the bamboo skewer, as in Fig. 6.4.

We support both sides of the needle on the back of a chair, above two cans or over another appropriate stand. We adjust the center of the bamboo skewer in relation to the center of the cork in such a way that the bamboo skewer remains horizontal. We then make two cuts on the upper part of the bamboo skewer, perpendicular to its axis, symmetrically located relative to the needle. The scale pans should be hung from these cuts. The friction of the needle rotating over the smooth stands is very small and this balance allows good precision. The needle works as the fulcrum of the balance, that is, the horizontal axis around which it can turn, as in Fig. 6.5.

Another way to reduce friction during the oscillation of the balance is to fix vertical nails or needles in the beam, which will be supported

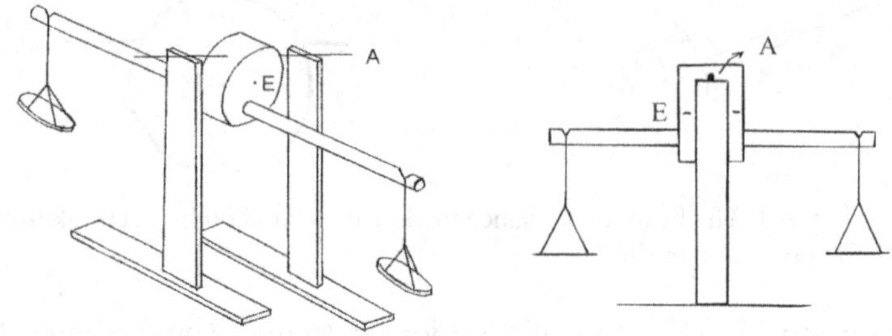

Fig. 6.5. A complete balance on its support.

over smooth stands. Fig. 6.6 illustrates a balance of this kind made with cork, bamboo barbecue skewers, and pins or needles.

Initially we use a knife to cut equal pieces of the cork from both ends, each one ⅓ of the length of the cork, in such a way as to remove ¾ of the circular part of the cork. Then we pass a bamboo barbecue skewer through the lower part of the cork, orthogonal to its axis of symmetry E. The bamboo skewer should be in a plane parallel to the plane of the longitudinal cuts of the cork, parallel to its axis of symmetry, but below the axis. We then remove its tip to make it symmetrical. Before passing the bamboo skewer through the cork we can pass a nail of the same thickness through the cork, in order to facilitate the insertion the bamboo skewer. We attach two vertical pins or needles in such a way that their tips are above the original axis of symmetry E of the cork. The bamboo barbecue skewer is set to horizontal, with the tips of the pins supported over appropriate stands of the same height. In order to stabilize the balance it is crucial that the tips of the pins be above the CG of the system composed of cork, pins and bamboo skewer. We can make small cuts close to the extremities of the bamboo skewer, perpendicular to the beam and on its upper side, in order to attach the threads for the scale pans.

There are several other possibilities, but what we have shown here should give a good idea how to build sensitive balances.

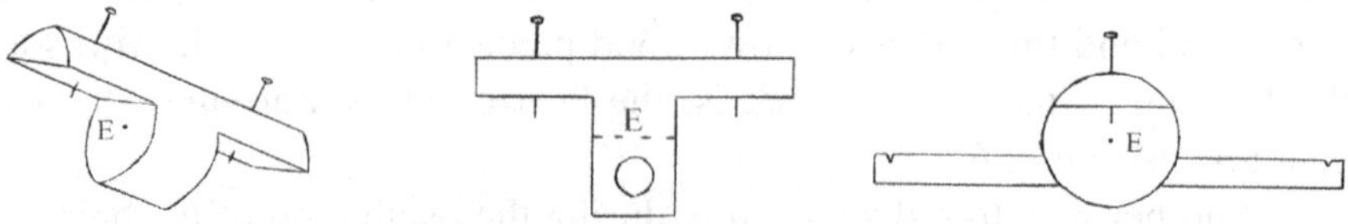

Fig. 6.6. Another kind of balance with very small friction.

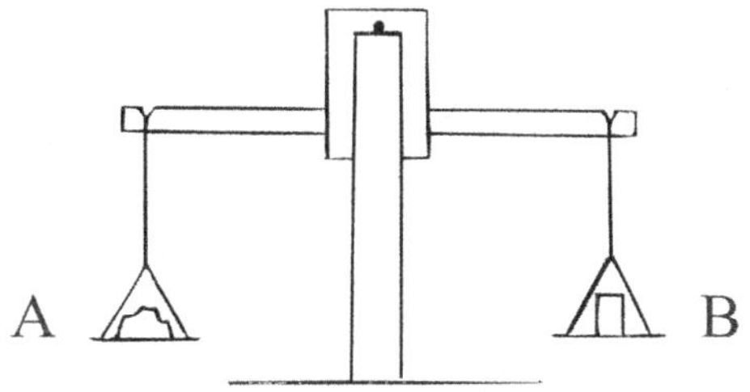

Fig. 6.7. A balance in equilibrium with equal weights.

6.2 Measurement of weight

We now show how to utilize a balance to weigh bodies. We will suppose that we have already built our equal arm balance and that it is completely free to turn around the fulcrum. Moreover, we will assume that it is in equilibrium; *i.e.*, the beam is horizontal without other bodies, only with the scale pans suspended at equal distances to the vertical plane passing through the fulcrum.

Experiment

We place body A (for example, a large paper clip) in the left pan of the balance and a sequence of N other bodies B (for example, a small paper clip, a large paper clip, a coin, a piece of modeling clay, *etc.*) in the right pan of the balance. In each trial we place only one of the N bodies B in the right pan, always releasing the balance from rest with its beam in a horizontal position. We observe that in some cases A goes up while B goes down, in other cases both bodies remain at rest with the beam horizontal, and in other cases A goes down while B goes up.

Definition: We say that bodies A and B have the same weight P if, when A is placed on one pan of this balance and B is placed on the other pan, and the beam is released from rest horizontally, it remains at rest, as in Fig. 6.7.

In order to obtain better precision with the balance, it is important to swap the position of the bodies on the scale pans. If it remains in equilibrium, we can say that the two bodies really have the same weight. There is one main reason for this precaution. It may happen that one of the arms (let us call it arm 1) is shorter than arm 2, this difference in lengths being difficult to detect with the naked eye. That is, the distance between the thread on arm 1 and the vertical plane passing through the fulcrum may be larger than the distance between the thread on arm 2 and the fulcrum. Let us suppose that these arms have different lengths. We further suppose that body A placed on pan 1 balances body B placed on pan 2. If the arms have different lengths, then body A placed on pan 2 will not balance body B placed on pan 1. The scale will only balance in both cases (A on pan 1 with B on pan 2; and A on pan 2 with B on pan 1) if the two arms are at the same distance from the vertical plane passing through the fulcrum. Switching the objects between scale pans is also necessary in the other cases where an equal arm balance is used. We will not mention this again, and simply suppose it is implied in other definitions and procedures.

We said before that body A (or B) has weight P, as if the weight belonged to it or were a property of the body A. However, as a matter of fact, the weight arises from an interaction of A with the Earth (or B with the Earth). We call this interaction gravity. The tendency of gravity is to unite the bodies with the Earth. Therefore, it would be more appropriate to say that when the balance remains in equilibrium, the interaction of A with the Earth has the same value P as the interaction of B with the Earth. In any case, we will keep the previous definition of the weight P of A and B, as this is the usual way of expressing it. But it should not be forgotten that weight is really an interaction of each body with the Earth.

The previous definition is an operational procedure for finding two bodies of the same weight. But it is not an experimental law. We are merely utilizing an empirical observation (the equilibrium of the balance supporting two bodies A and B) in order to arrive at a conceptual (or operational) definition.

This could only be an experimental law if we had some other way of knowing when two bodies have the same weight. If this were the case, then we could say that experiment teaches us that two bodies of the same weight keep an equal arm balance in equilibrium. But historically it was with the equal arm balance that we first found an objective procedure for quantifying the notion of weight. Therefore, the equality of weight of two bodies by this first procedure must be established by definition. Once we know one procedure for defining the equality of weight of two bodies, we can apply it to obtain other experimental laws. For instance, suppose we utilize the previous experimental procedure with an equal arm balance to find two bodies A and B of equal weight. Then we can raise to an experimental law the empirical result that these two bodies compress a spring by the same amount when each one of them remains at rest above the ground and above a vertical spring supported from below.

The previous definition is the main operational procedure for quantifying the equality of weight of two bodies. We might think of an alternative procedure such as: we define two bodies made of the same material and having the same size and shape as having the same weight. But this alternative procedure has problems and limitations, for two principal reasons. The first is that it is difficult to know in practice if the two bodies are really made of the same material. After all, microscopic differences may arise during the manufacturing process (impurities, internal bubbles, *etc.*) which are difficult to detect. Even disregarding this prospect, there is a second, even more serious problem. There is not the slightest possibility of comparing the weights of two bodies made of different materials, such as iron and wood, or corn and water, by this alternative definition. That is, when we have bodies of different chemical composition, we cannot compare their weights by this alternative definition.

Let us illustrate this point with a specific example, as this is a relevant issue that is rarely discussed in textbooks. When we buy a box of paper clips we observe visually that they have the same shape and size. As they are made of the same material, it is reasonable to suppose that they have the same weight. Despite this fact, there are always some microscopic variations between two clips which are difficult to detect macroscopically. In any event, even forgetting this fact, there is not the slightest possibility of visually comparing the weight of one of these clips with a certain amount of clay. After all, the clip and the clay have different shapes, sizes, volumes, textures, colors, *etc.* But the main difference is that they are made of different chemical substances. The only way of knowing if they have the same weight or not is to utilize a measurable effect arising from the gravitational interaction. The first quantitative instrument that was devised historically to determine the weight of bodies was the equal arm balance. Therefore, we say, by definition, that a clip and a certain volume of clay have the same weight if, when released at rest on the pans of the balance, the beam remains horizontally at rest.

In principle the previous definition is only strictly valid when the balance is placed in a high vacuum. The reason is that if bodies A and B are immersed in a fluid like air, an upward force will be exerted upon them by the air. And this force is equal to the weight of the displaced air, as discovered by Archimedes himself. Therefore, the body with the larger volume will receive a larger upward force from the air. This force of the air will distort the comparison of weights of A and B. In our previous definition we are neglecting the effect of this upward force, consid-

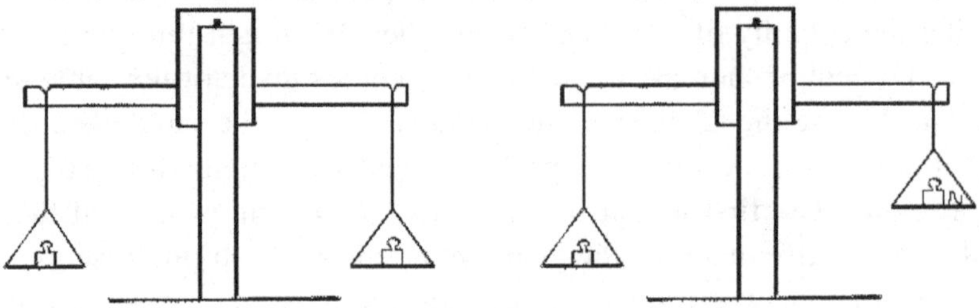

Fig. 6.8. The weight does not depend upon the height of the body.

ering only the downward forces upon A and B due to their interaction with the Earth.

Definition: We place two bodies A and B on different pans of a equal arm balance, with the beam initially horizontal, releasing the system from rest. If the balance does not remain in equilibrium, but inclines towards one of the sides, we say that the body which moves toward (away from) the ground is the heavier (lighter).

Experiment

We utilize an equal arm balance with threads of equal length holding the two pans. We find two bodies A and B which keep the balance in equilibrium. We then shorten one of the threads, placing the excess thread on the pan to which it belonged, and again release the beam from rest with bodies A and B on the two pans. We observe that the balance remains in equilibrium as in Fig. 6.8. In other words, experimentally the weight of a body does not depend on its height above the ground.

Since Newton's theory of universal gravitation we have known that this result is only an approximation, since the gravitational force between two spherical bodies falls as the inverse square of the distance between their centers. But due to the huge radius of the Earth, compared with the difference in length between the two threads in this experiment, the change of weight is negligible. Thus, it cannot be detected in this kind of experiment. We can therefore assume as an experimental result that the weight of a body upon the surface of the Earth does not depend upon its height above the ground.

Now that we have defined the equality and inequality of weight between two bodies, we can continue to quantify the notion of weight with another definition.

Definition: We say that N bodies of the same weight placed together on a scale pan have N times the weight of one of these bodies.

For example, suppose that with a equal arm balance we discover that the bodies A, B, C and D all have the same weight P ($P_A = P_B = P_C = P_D \equiv P$). Suppose we place these four bodies over one of the pans of a balance and verify that they equilibrate another body E placed on the other pan. Then we say, by definition, that the weight of E is four times the weight of A, or, $P_E \equiv 4P_A$.

This may seem a trivial definition. But this is not the case. In order to see this, let us compare it with the temperature of a body. We define two bodies as having the same temperature T if, when they are placed in contact, they remain in thermal equilibrium. That is, their macroscopic variables, like the pressure or volume in the case of gases, do not change with the passage of time. But if we place N bodies together at the same temperature T, the system as a whole will also have the same temperature T, and not a temperature N times higher than T. The same holds for density. That is, when we place N cubic solid homogeneous bodies of the same density ρ together, the system as a whole will have the same density ρ. The system will not have N times this density.

Based on this definition we can prepare a set of standard weights. We choose a specific object, such as a small paper clip, as our standard, and define its weight as 1. With a balance we can find many other objects (*e.g.*, pieces of clay) which have the same weight. We then put five of these equal weights on one side of a balance and on the other side we place an appropriate amount of clay to balance these 5 objects. This clay will have, by definition, weight 5. We can mark this number in the clay. We can find other standards of weight: 10, 50 and 100, for instance. Now suppose we want to weigh an apple. We put it on one side of the balance and find how many units we need to place on the other side to balance it. If it is 327 units, we will say that the weight of the apple is the same as the weight of 327 paper clips, or simply 327 units.

We now present a series of experiments that show how to improve the sensitivity of balances.

6.3 Improving balance sensitivity

We now perform four experiments. Their results will show how to build balances with greater sensitivity.[2] All of them utilize pasteboard figures in the shape of the letter T, as in Fig. 6.3. This pierced pasteboard T will

[2] N. Ferreira, *Equilíbrio* (São Paulo: Instituto de Física, USP). Available in 2007 at: http://www.ludoteca.if.usp.br/.

function as a balance. Its arms of equal length will be the beam of the balance. We will suppose that when we hang the T by the hole located at the intersection of the arms with its body, supporting it on a horizontal pin fixed in a vertical stand, the arms of the balance remain horizontal after the T stops swaying. We then find two bodies that keep the T in equilibrium when they are placed on opposite sides at equal distances from the axis of symmetry of the T. The balance is being utilized here to determine the equality of weight of these two bodies. But we can also use a balance to determine whether two bodies A and B have different weights. How should we build a balance capable of distinguishing, for instance, a difference in weight of 1% between A and B? We are interested here in finding the main features that increase the sensitivity of a balance, so that it can easily show that two bodies A and B have different weights. This is the goal of the next experiments.

Suppose we hang two bodies A and B, having different weights, from opposite arms of a balance 1 and a balance 2. We will say that balance 1 has a higher sensitivity than balance 2 if we can more easily distinguish the difference of weight in balance 1 than in balance 2. The sensitivity of a balance can be established quantitatively by the angle θ its arms make with the horizontal when it holds bodies A and B at equal distances to its fulcrum. The greater the value of θ, the greater the sensitivity of the balance.

In order to unbalance the beam we will use a paper clip placed on one of its arms. We want to know what makes the disequilibrium more visible, that is, what increases the angle θ indicated by the T.

The dimensions of the T do not need to be exactly as indicated. In the model used here, the length between the end of one arm and the end of the other arm is 15 cm. The height of the T is 16.5 cm. The width of the arms and body of the T is 3 cm. Holes separated by 1.5 cm are made along the axis of the arms and along the axis of symmetry of the T. Let us call the 10 holes along the axis of symmetry V_1 to V_{10}, with V_1 at the intersection of the arms with the body, and V_{10} at the bottom end of the body. The holes along the arms are called H_1 to H_8, with H_1 at the left of Fig. 6.3 and H_8 at the right.

After making these holes we locate the CG of the T. The simplest way to do this is to hang it by a pin passing through H_1, and draw the vertical with the help of a plumb line after the system has reached equilibrium. This procedure is repeated with the T hanging by H_8. The intersection of the two verticals is the CG of the T. With the previous dimensions it is located between V_3 and V_4, as indicated in Fig. 6.3.

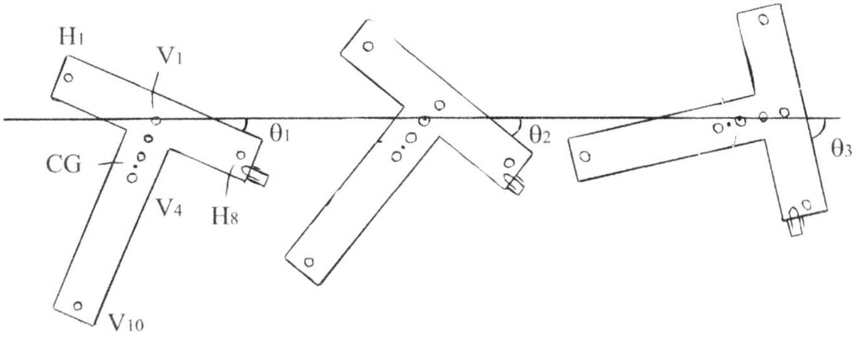

Fig. 6.9. The greater the distance between the point of suspension PS and the CG, the smaller the sensitivity of the balance.

Experiment

Initially we have a balance in equilibrium, with its arms horizontal, suspended by hole V_1. We now disturb this equilibrium by placing a small piece of paper or clay, or a paper clip, at the end of one of the arms. The system turns around V_1, oscillates a few times, then stops with the extra weight lower than the opposite arm. Let us call the smaller angle between the horizontal and the arm with the extra weight when the T is at rest θ_1, as in Fig. 6.9. We repeat the experiment, but now with the T suspended by V_2. Initially the system is in equilibrium with the arms horizontal. We then disturb this equilibrium by placing the same extra weight in the same place as before. After the system has come to rest we measure the angle between the horizontal and the arm with the weight, calling it θ_2. We repeat the procedure with the T suspended by V_3. In this case, the angle when the system is at rest is called θ_3. Experimentally it is found that the smaller the distance between the point of suspension (in this case, the pin) and the CG of the T, the greater the final angle when the system is at rest. That is, experiment shows that $\theta_1 < \theta_2 < \theta_3$.

If we try to keep the T in its normal position (with the arms above the body) by suspending it by holes which are below the CG, we do not succeed. In other words, if we try to suspend it by V_4, V_5, ..., V_{10}, the system turns and remains at rest only with the horizontal arms below the body of the T. But even in these cases we can break the equilibrium as before, and obtain the same experimental results. That is, if we suspend the T by V_{10} and place an extra weight at the end of one of its arms, the system will reach a new position at rest with the arm inclined by an angle θ_{10} from the horizontal, as in Fig. 6.10. We now hang the T by V_9, ..., V_4, then put the same extra weight at the end of its arm, and wait until the system reaches equilibrium. In these cases the smaller angle between

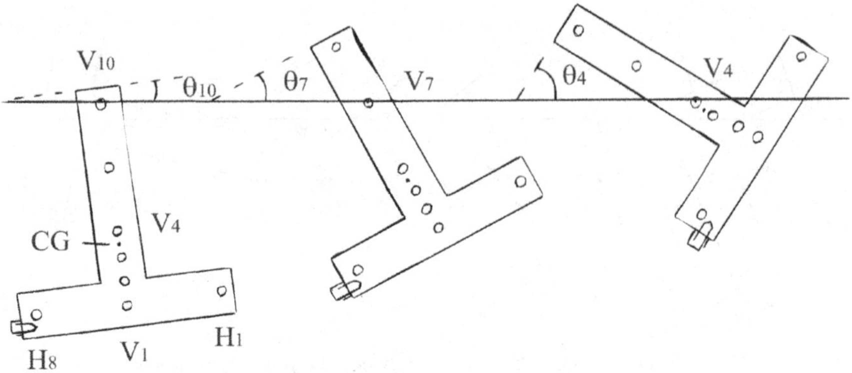

Fig. 6.10. The same result as before with the T upside down.

the horizontal and the arm with the extra weight is given by θ_9, ..., θ_4, respectively. Experimentally it is found that $\theta_{10} < \theta_9 < ... < \theta_4$.

In all these cases we placed the same weight acting at the same distance from the vertical plane passing through the fulcrum of the balance. And we discovered experimentally that the smaller the distance between the PS and the CG, the greater the angle of inclination of the beam with the horizontal after the system reached rest. Therefore, the sensitivity of a balance increases with decreasing distance between the PS and the CG. As the distance between the PS and the CG gets smaller, it is easier to perceive that the beam is unbalanced, supporting different weights on both arms.

This experiment suggests that balances should be built to allow a variable distance between the PS and the CG, in order to control sensitivity. An example of a balance of this kind utilizes a cork, two bamboo barbecue skewers and two pins or needles. Initially we pass a bamboo skewer through the cork, orthogonally to its axis, at a distance of ⅓ of its length from one end. We remove the tip of the bamboo skewer and make two cuts on the upper face of the bamboo skewer, at the same distance from its center, in order to support the threads fixed to the scale pans. We then pass another bamboo skewer at a distance of ⅓ of its length from the other end, in such a way that it remains orthogonal to the axis of the cork and the first bamboo skewer. This second bamboo skewer will work as the pointer of the balance. We place a pin parallel to this second bamboo skewer, passing close to the center of the cork, to serve as the fulcrum of the balance. In order to prevent the beam from falling toward the ground when we place the threads and scale pans, raising the pointer, we place another pin parallel to the first one, this time in the front part of the cork, after the horizontal bamboo skewer.

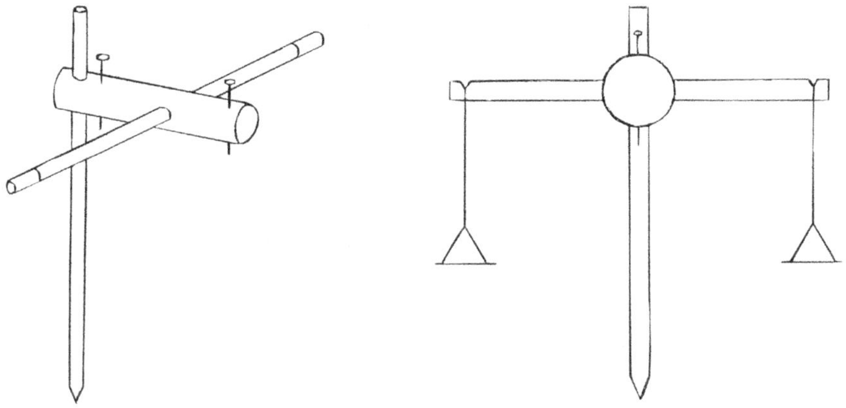

Fig. 6.11. A balance with variable distance between the PS and the CG.

We then have along the length of the cork, from back to front: a vertical pointer, a vertical pin, the horizontal beam and another vertical pin, as in Fig. 6.11. We fix the two scale pans and adjust the arms so that the beam becomes horizontal when supported by the two pins. We then support the balance with the two pins on the lid of a can or other convenient support. The balance is then complete. By raising or lowering the vertical bamboo skewer we can change the height of the CG of the balance. In this way we can change its sensitivity, as desired. This vertical bamboo skewer works as well as the pointer of the balance. For example, when the balance is equilibrated with its arms horizontal, we can make a small mark on the stand parallel to the location of the pointer indicating the zero (0) of the balance.

Another extremely creative idea to connect two bamboo skewers or two plastic straws, without a cork, is to make a loop out of pieces of a plastic straw.[3] To do this, we cut three small pieces of straw, one 4 cm in length and two 5 cm in length. The larger pieces are folded in two and we introduce them into the smaller piece. The angle between the planes of the two loops should be 90°, as in Fig. 6.12. We pass a whole straw or bamboo skewer through each loop and stick two pins or needles in the 4 cm long straw. The two bamboo skewers or whole straws should be orthogonal to one another. The bamboo skewer parallel to the two pins or needles will be the pointer of the balance. In this way we can support the two pins over a rigid stand. The horizontal bamboo skewer will be the beam of the balance. The length of its two arms should be adjusted with the beam remaining horizontal at rest. After this we draw one mark on each arm at equal distances from the point of intersection. On these

[3] N. Ferreira, *Equilíbrio.*

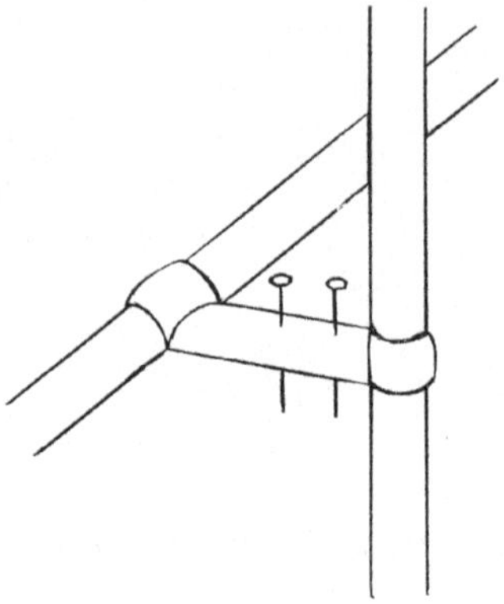

Fig. 6.12. A balance with variable distance between the PS and the CG made of plastic straws.

marks we hang the threads with scale pans. The vertical bamboo skewer (the pointer) can be adjusted at will, so that we can change the distance between the points of suspension (lower tip of the pins) and the CG of the system (composed of bamboo skewers, pieces of straw, pins and scale pans). In this way we can control the sensitivity of the balance. In order to prevent the balance from falling when we put objects on the scale pans, the objects should be very light, with a weight no larger than the weight of the system. If we wish to balance heavier bodies, then we will need to put extra weights over the pointer in order to prevent the whole balance from falling.

It is important to improve the sensitivity of a balance. But this has a side effect. When we remove a balance from its position of stable equilibrium and release it from rest, it oscillates a few times until it stops due to friction, returning to its position of stable equilibrium. But the smaller the distance between the PS and the CG, the longer will be the period of oscillation. It will therefore take a longer time for the balance to complete each oscillation. When the PS is very close to the CG, we sometimes need to wait a long time until the balance stops swinging. This creates problems because it takes a long time to make each reading of the balance. This makes certain measurements impractical, as small perturbations in the position of the beam are inevitable (due to air currents, tremors of the room, perturbations when we place bodies over the scale pans, *etc.*) To prevent this problem some balances have a damper or shock-absorber (such as a pointer inside a vessel of oil) which quickly decreases the amplitude of oscillations. In this way we can move the CG close to the PS, increasing the sensitivity of the balance,

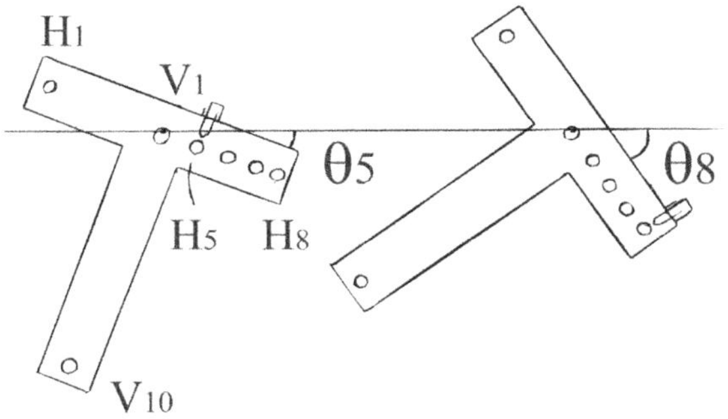

Fig. 6.13. The longer the arms of a balance, the greater its sensitivity.

without significantly increasing the time for each reading of the balance due to perturbation.

In the next experiment we analyze another effect which shows how to increase the sensitivity of a balance.

Experiment

In this experiment we always hang the T by the same hole, such as V_1. Let us suppose that it remains at rest in this position with its arms horizontal. We now disturb this equilibrium by placing an extra weight (a piece of paper or clay, or a paper clip) over the hole H_8, releasing the system from rest. The T oscillates a few times, stopping with H_8 below H_1. Let θ_8 be the smaller angle between the horizontal and the beam in this final position. We now remove the extra weight from H_8, and place it over H_7, releasing the beam from rest in a horizontal position. After a few oscillations the system stops with H_7 below H_1. Let θ_7 be the smaller angle between the beam and the horizontal in this final position. The procedure is repeated with the extra weight over H_6 and over H_5. These experiments show that $\theta_8 > \theta_7 > \theta_6 > \theta_5$, as in Fig. 6.13.

We can imagine that in these four situations we have the same balance, but with the scale pans hanging by equal arms of different lengths in each instance (by H_1 and H_8 in one situation, by H_2 and H_7 in another situation, by H_3 and H_6 in another situation, and by H_4 and H_5 in another situation). We conclude that the longer the arms of a balance, the greater its sensitivity. That is, by comparing two balances with the same distance between the PS and CG, the more sensitive balance is the one with the longer arms. After all, the longer the arm with the extra weight, the more visible will be the lack of equilibrium of this balance caused by objects A and B of different weight. This lack of equilibrium is indicated by the angle of inclination of the beam with the horizontal.

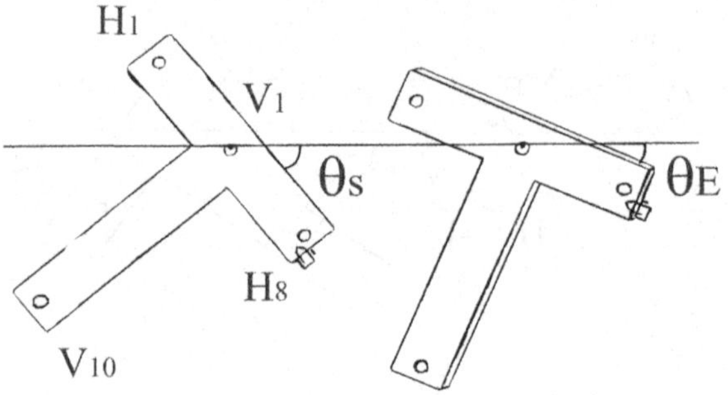

Fig. 6.14. The lighter a balance is, the greater its sensitivity.

The results of these two experiments can be combined in a single expression. Let h be the vertical distance between the PS and the CG of the balance. Let d be the arm of the balance (horizontal distance between the point of suspension of the scale pans and the vertical plane passing through the fulcrum). The larger the ratio d/h, the greater the sensitivity of the balance. Or the larger the angle θ of inclination of the beam to the horizontal when there are different weights on the scale pans.

Experiment

A third effect that illustrates how to improve the sensitivity of a balance can also be easily seen with a pasteboard T. In this case we cut out three or four equal T figures, of the same size and shape. Two or three of them should be glued together, making a T of the same size as the original one, but now two or three times thicker than a single T. The two systems (the single T and the thick T) have holes in the same locations (V_1 to V_{10} and H_1 to H_8). We can determine the CG of both systems experimentally. They coincide with one another, being located between holes V_3 and V_4. We hang the single T by V_1 and wait until the system reaches equilibrium with its arms horizontal. We then suspend an extra weight, like a paper clip, at the extremity of one of its arms. We wait until the system stops its oscillations, with the arm containing the extra weight lower than the other arm. We measure the angle θ_S between the horizontal and this arm. We remove the T from the support and hang the thick T by V_1. We suspend the same extra weight at the end of one of its arms. We wait for the system to stop moving and measure the angle θ_E between the horizontal and this arm. Experimentally we observe that $\theta_S > \theta_E$, *i.e.*, the heavier the beam of a balance in comparison with the extra weight, the less sensitive it will be. In this experiment the distance between the PS and the CG of the balance was the same, and the extra weight always hung at the same distance from the vertical passing

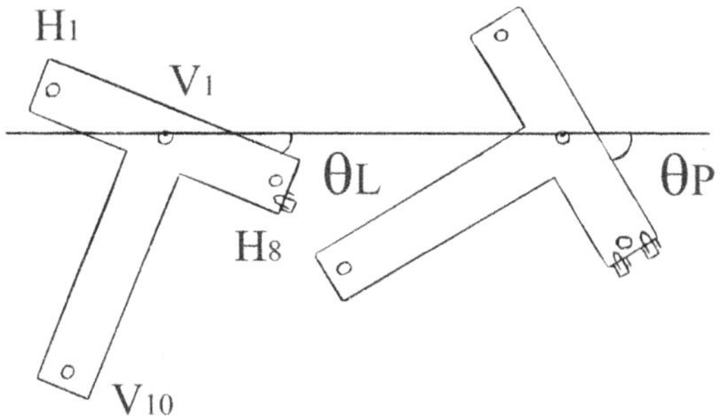

Fig. 6.15. The greater the difference in weight between the two bodies, the more easily we will notice it.

through the fulcrum. The different sensitivity of the two balances can only be due to the difference in their weights. We conclude that the lighter a balance is; the more sensitive it will be to distinguish the same difference of weight between two bodies, as illustrated in Fig. 6.14.

Experiment

It is also easy to observe experimentally that the greater the extra weight placed upon one of the arms of a balance, the more inclined the beam will be from the horizontal. For example, we hang an extra weight upon one of the arms of a balance and wait until the system stops its oscillations. Let θ_L be the angle between the horizontal and this arm. We now place two extra weights upon the same arm, at the same distance from the fulcrum. Once again, we release the balance from rest, with its beam horizontal, waiting until it stops its oscillations. Let θ_P be the new angle between the horizontal and this arm. Experimentally it is found that $\theta_L < \theta_P$, as in Fig. 6.15. This means that the greater the difference in weight between the bodies on the two equal arms of the balance, the more easily we will notice it, or, the greater the final angle of inclination between the beam and the horizontal.

Once more we can combine the results of these last two experiments in a single expression. Let $\Delta P \equiv |P_A - P_B|$ be the magnitude of the difference of weight between A and B. Let us represent the weight of the beam by P_A. Therefore, the greater the value of $\Delta P/P_A$, the greater will be the sensitivity of the balance, or, the greater the angle θ of inclination of the beam from the horizontal when ΔP is different from zero. If ΔP is the same for two different balances, the balance with a lighter beam will be more sensitive.

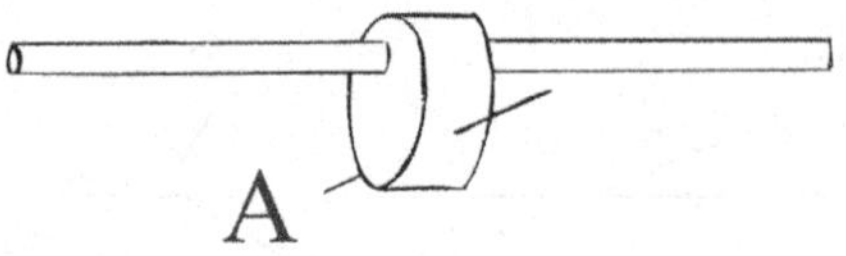

Fig. 6.16. A beam with its CG above the fulcrum (needle A).

6.4 Condition of equilibrium of a suspended body

Before studying levers it is worth making another experimental observation. Let us consider the balance with bamboo skewer, needle (A) and cork, where the axes of symmetry of these three bodies are in the horizontal position, as in Figs. 6.4 and 6.5.

Experiment

The balance is in stable equilibrium when the needle is above the center of the cork and above the center of the bamboo skewer, with or without the scale pans. That is, when we lower one of the sides of the bamboo skewer and release it from rest, the balance sways a few times, stopping with its arms horizontal (supposing there are equal weights on its scale pans suspended at equal distances from the fulcrum). It is easy to understand this fact by observing that in the position of stable equilibrium the CG of the system is in its lowest possible position, below the needle, along the vertical line passing through the center of the needle. Any perturbation raises the CG. Therefore, if the system is free to rotate after release, it will return to the position of stable equilibrium.

Experiment

We now consider the opposite case in which the center of the needle is below the center of the cork and below the center of the bamboo skewer. Let us suppose initially that there are no scale pans on the beam, as in Fig. 6.16. In this case the equilibrium is unstable with the horizontal needle. In other words, we cannot keep the balance at rest in this position after release; it tends to turn in the clockwise or in the anticlockwise direction after being released from rest. If the balance can make a complete turn, it will end up in the previous position of stable equilibrium. It is also easy to understand the phenomenon by observing that in the position of unstable equilibrium the CG of the system is in its highest possible position, above the needle, along the vertical line passing through the center of the needle. Any perturbation in the system tends to lower its CG. Therefore, if the balance begins to turn in the clockwise direction

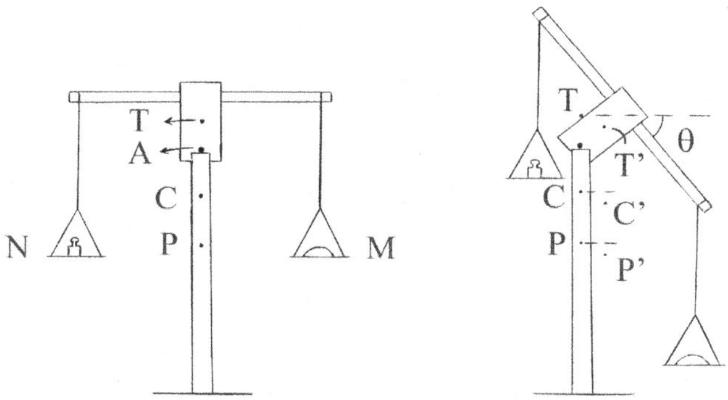

Fig. 6.17. A balance in unstable equilibrium.

after being released from rest, it will continue to turn in this direction, as the tendency of the CG is to fall toward the ground.

Experiment

The most curious situation is when the center of the needle is in the previous position, below the center of the cork and below the center of the bamboo skewer, but now with equal weights M and N placed on arms of equal length. Let us suppose that the balance has the bamboo skewer (the beam) initially horizontal. Moreover, let us suppose that the weight of the set of threads and scale pans, together with objects M and N placed on these pans (CG of this first set located at P) is larger than the weight of the set of cork, needle and bamboo skewer (CG of this second set located at T), in such a way that the CG of both systems together is located at C, below the needle A, as in Fig. 6.17. Even in this case the system is in unstable equilibrium in this initial configuration. That is, if released from rest it tends to turn in the clockwise or in anticlockwise direction. The beam of the balance does not remain in this initial position if there is any perturbation in the system.

Let us try to understand what is happening here. We first analyze the situation for which the beam has turned an angle θ from the horizontal, in such a way that body M moves downward and body N upward, as in Fig. 6.17. Body M, together with its pan and thread, fell a distance $H(\theta)$ relative to its original height above the ground. At the same time body N rose, together with its pan and thread, a distance $h(\theta)$ relative to its original height above the ground. As the center of the cork also fell below original height, we have $H(\theta) > h(\theta)$. This means that the CG of the first set (bodies M and N, together with their plates and threads) fall relative to its original height above the ground, from P to P′. The same happened with the CG of the second set (cork, needle and bamboo skewer), moving from T to T′, and with the CG of the whole system, which moved from C to C′. This means that the tendency of the system will be to increase the angle θ even more, as this will lower the CG of the whole system.

If the system had turned an angle θ relative to the horizontal in such a way that N went downwards and M upwards, the CG of the whole system would have again moved downward relative to its original position. And the system would tend to increase angle θ even more. And this explains the unstable equilibrium in this case.

We call attention to this case because it brings something new. When we were considering the equilibrium of rigid bodies, we could only obtain unstable equilibria with the CG above the point of support PA (as, for instance, the case of the parallelogram turning around one of its edges). This happened when any perturbation in the position of the body lowered its CG. On the other hand, we had seen stable equilibrium with the CG above the PA (as, for instance, a rocking chair oscillating on a flat surface). We had also seen stable equilibrium with the CG below the PS (plane figures suspended by a needle passing through one of their holes). In these latter two cases the stable equilibria arose when any perturbation in the position of the body raised its CG.

In the present case we no longer have a rigid body. When the beam turns by an angle θ relative to the horizon, the angle between the beam and the threads supporting the scale pans is modified (it is no longer a right angle). Moreover, the distance between the center of each pan and the center of the beam has also been changed. We are now seeing a new kind of unstable equilibrium, a case where the CG of the whole system is below the PS. And we again conclude, but now in a more general situation not restricted to rigid bodies, that there will be stable (unstable) equilibrium whenever the CG of the whole system rises (falls) when there is any perturbation in the configuration of the system. There will be neutral equilibrium when the CG of the system remains at the same height for any perturbation of the system.

The key to obtaining stable equilibrium of a balance which is free to turn around a horizontal axis is that the PS should be located vertically above the CG of the beam. We mentioned this earlier, but it is important to emphasize it here once more. For example, if the beam is a rectangular beam of wood or a cylindrical rod, the fulcrum should not be placed at the center of the beam or cylinder. In order to obtain stable equilibrium, the fulcrum or PS of the balance should be located above the center of the beam. This will guarantee the stability of the balance when it is placed with its beam horizontal. If the fulcrum is placed ex-

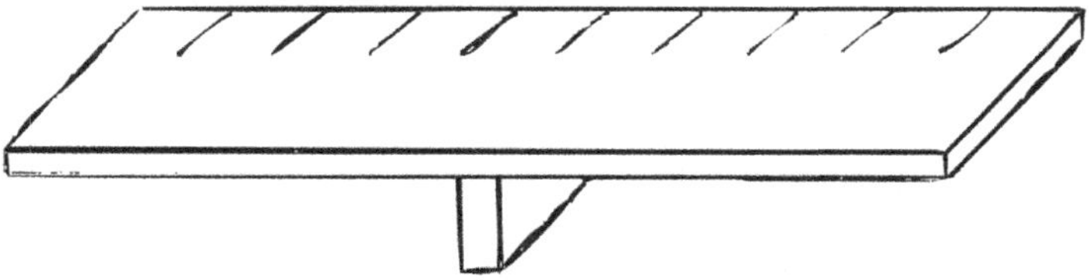

Fig. 6.18. A balance with its CG above the fulcrum.

actly at the center of the beam, a procedure that will produce stable equilibrium is to fix an extra weight on the beam, located vertically below the fulcrum. This will lower the CG of the beam, in such a way that the new CG will be lower than the fulcrum (or PS).

6.5 Balances with the center of gravity above the fulcrum

Before moving on, we briefly mention balances which have the CG of the beam above the fulcrum. As there is unstable equilibrium in this case, the only way to build a working balance is to support it on a surface, not on a point or single horizontal line without thickness. An example of a balance of this kind is a horizontal ruler supported by a domino piece placed below its center, as in Fig. 6.18. The ruler can only remain at rest if the width of the domino touching the ruler is not too small in comparison with the thickness of the ruler. For example, it is extremely difficult to balance a horizontal ruler on the edge of a vertical razor blade. In this case the ruler falls to one side or another even before we put the weights on it.

This setup limits the precision or sensitivity of the balance. After all, the surface on which the beam is supported does not allow a single distance between the weights above the pans and the vertical plane passing through the fulcrum. The distance of each arm from the vertical plane passing through the fulcrum can take any value between a minimum and a maximum. As a result, with this apparatus we can balance bodies of the same weight and bodies of different weights (as established by the precise balances already presented, for which the fulcrum was above the CG of the beam).

Another problem with these balances is that the supports for the weights to be measured (small cups, bottle caps, *etc.*) are normally attached to the beam. Therefore, the weights are not supported by a single point, as they are spread over a small region. This is another reason why it is difficult to find a single distance between each arm (or each weight) and the vertical passing through the fulcrum.

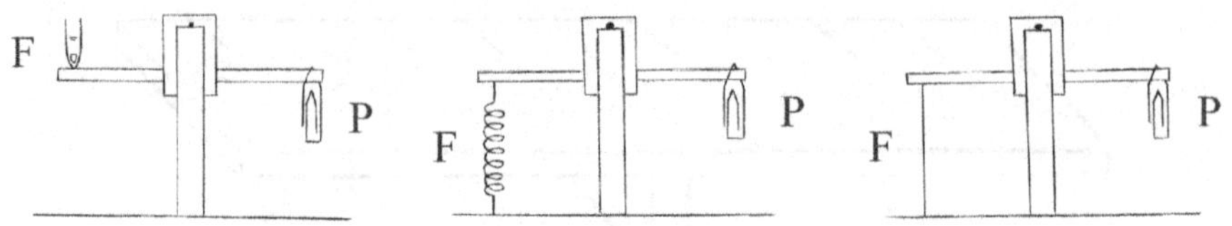

Fig. 6.19. Utilizing the weight as a force standard.

6.6 Other types of balance

Apart from the equal arm balance there are other types which utilize other measurable effects due to the action of gravity. A common balance for home use is made of springs. It utilizes the compression of a spring due to a body at rest in a pan as a weight indicator. Some high-precision piezoelectric balances utilize a phenomenon observed in anisotropic crystals as a weight indicator. Some crystals, when mechanically compressed, become electrically polarized in certain directions. This can be measured and calibrated to indicate the weight compressing the crystal. Some electronic balances transform mechanical deformations arising from the weight of a body into electrical voltage, which is measured electronically. There are several other kinds of balance, but we will not consider them here.

6.7 Using weight as a standard of force

It is possible to keep the beam of an equal arm balance horizontal by placing a body of weight P on one side, while on the other side, at the same distance from the vertical plane passing through the fulcrum, another mechanism sets the balance. In order to simplify the analysis we will suppose that the balance has no scale pans, in such a way that the weight P is suspended directly by the beam. The mechanism which counterbalances the weight P can be, for instance, the finger of a person exerting a downward force. It can also be a tensed spring fixed at the

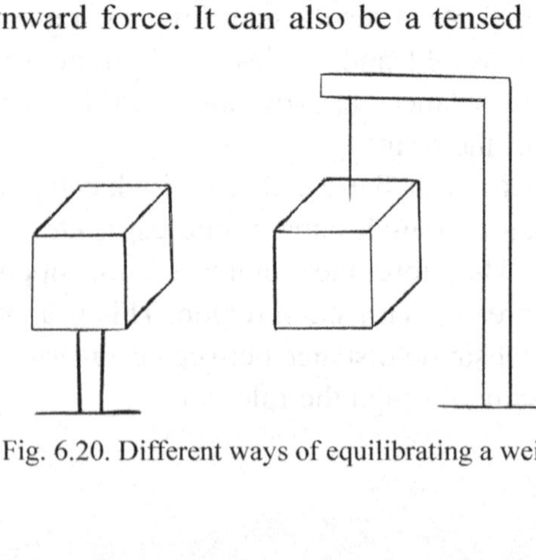

Fig. 6.20. Different ways of equilibrating a weight.

ground below the balance, or a taught thread fixed to the ground, as in Fig. 6.19. Several other mechanisms can operate on the other side to the weight P in order to equilibrate it (mechanisms depending upon electric or magnetic effects, for instance). This leads to an important definition.

Definition: Suppose body A of weight P acting at a distance *d* from the fulcrum of a balance, being equilibrated by a second body B acting on the other side of the balance, at the same distance *d* from the fulcrum. We define that this second body B exerts a force of magnitude F equal to the weight P of the body A, regardless of the nature of this force (it can be elastic, electrical, magnetic, *etc.*) That is, $F \equiv P$.

In this case we stipulate that the finger (or spring, or thread, or magnet, *etc.*) exerts a force of magnitude F equal to the weight P of the body. As a result, we can calibrate or measure forces of different kinds, not necessarily gravitational, by comparing them quantitatively with the force due to the weight.

This concept does need not be limited to an equal arm balance. We have seen that when we release a body from rest above the surface of the Earth, it falls to the ground. But this can be prevented by different means, for instance, by placing a rigid support or spring under the body, or suspending it by a thread or spring, *etc.* Fig. 6.20 illustrates a few possibilities.

Let us consider a spring at rest vertically, fixed at its upper end, with a total length L_0 in this vertical position, as in Fig. 6.21. When a body of weight P is suspended and kept at rest at the lower end of this spring, the spring acquires a length $L_1 > L_0$. Another way to keep this body at rest relative to the ground is to support it on the upper end of a vertical spring, which has its bottom end fixed on the ground. In this case, the spring is compressed to a length $L_2 < L_0$. By definition we say that in these cases the tensed or compressed spring exerts an upward force F upon the body of weight P given by $F \equiv P$. This also holds if, instead of the spring, the body is suspended by a thread, supported by a

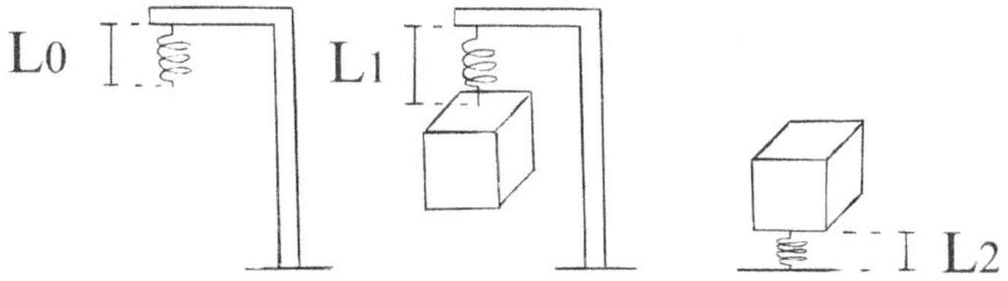

Fig. 6.21. A stretched or compressed spring balancing a weight.

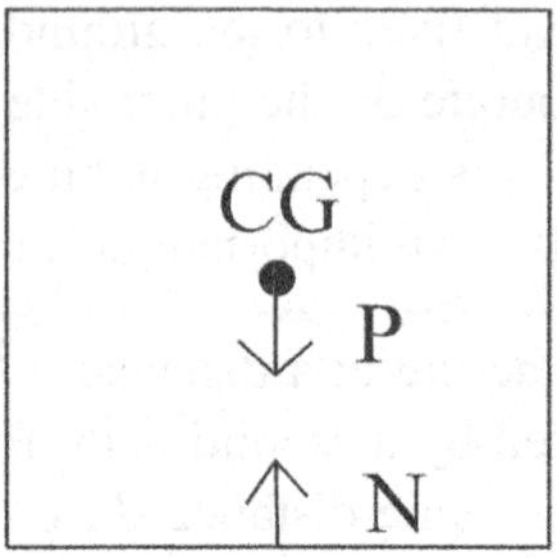

Fig. 6.22. The weight of the body equilibrated by the normal force exerted by the pan.

stick or a person's hands, *etc.*

We have seen that if an object A is released from rest, it falls to the ground. In the previous experiments we saw that we can prevent this by connecting this body to an equal arm balance and placing another body B on the other side of the balance. We define that these two bodies have the same weight if the balance remains in equilibrium. But body A is not connected directly to body B, as it is in contact only with the pan of the balance. We can then see that the downward weight acting upon A, due to the Earth's gravity and acting as if it were concentrated at the CG of A, is balanced by a normal upward force of magnitude N exerted by the pan of the balance acting upon A at the region of contact. That is, $N \equiv P$, as in Fig. 6.22. This normal force N has its origin in the downward weight of body B, being transmitted by the curved pan and taught thread holding B, by the curved rigid beam, and then by the taught thread and curved pan holding A. The threads holding the scale pans are taught (that is, under tension) due to the gravity acting upon A and B. The scale pans are also under stress or tension, with the threads forcing them upward, while A and B force them downward. We can then say that a first condition of equilibrium in order for a body to remain at rest relative to the ground is that the downward weight P must be counterbalanced by an upward force N of the same magnitude as the weight.

We can also investigate weight and forces in general by considering algebraic magnitudes, that is, positive and negative. We deal here with forces along the vertical direction and choose the downward direction as positive. In other words, forces exerted toward the Earth, such as the weight, are considered positive, while upward acting forces are considered negative. We can also choose, for instance, the right and forward directions as positive, while the left and backward directions will be negative. We then postulate that a body is in equilibrium when the sum of all forces acting upon it, in all directions, goes to zero. If this sum is different from zero, we postulate that the body will move toward the direction of the net force.

Chapter 7

Levers and the First Law of Mechanics

7.1 Building and calibrating levers

The lever is one of the simple machines studied in ancient Greece. The other simple machines were the windlass, the pulley, the wedge and the screw. The lever consists of a rigid body, normally linear, the beam, capable of turning around a fixed axis horizontal to the ground. This axis is called the fulcrum or point of suspension, PS, of the lever. This axis is orthogonal to the beam. The lever is like a balance, but now with the possibility of placing weights at different distances from the fulcrum. The models which we will consider here are analogous to the balances built earlier. We will consider only levers in stable equilibrium for which the fulcrum is vertically above the CG of the beam when it is at rest horizontally. We will suppose that the lever is symmetrical about the vertical plane passing through the fulcrum, with the beam horizontal and orthogonal to this vertical plane when there are no bodies supported by the lever. As we did with the balance, we will define the expression "lever in equilibrium" when its beam remains at rest horizontally relative to the ground. We call the arm of the lever the horizontal distance d between the point of suspension of a body upon the beam and the vertical plane passing through the fulcrum. For brevity we sometimes say, simply, "distance between the body and the fulcrum;" but in general this should be understood as meaning the horizontal distance between the point of suspension of the body upon the beam and the vertical plane passing through the fulcrum. When we talk about the two arms of a lever, these should be understood as the opposite sides in relation to the vertical plane passing through the fulcrum.

In order to arrive at the first law of mechanics in a precise and quantitative way we need a sensitive lever. The conditions to obtain this are the same as for the balance: freedom of rotation around the fulcrum; a high ratio $\Delta P/P_A$ (where ΔP is the difference of weight between the bodies suspended on the two sides of the lever, and P_A is the weight of the lever); and a high ratio d/h (where h is the vertical distance between the PS and the CG of the beam, while d is the smaller arm of the lever).

We also need to mark precisely upon the two arms several points at equal distances from the vertical plane passing through the fulcrum. There are two ways to do this. (A) The first is to establish the fulcrum of the lever (by making a hole or attaching a hook from which it will hang; or passing a needle through the beam, in such a way that it is attached to the beam so it can be supported over a stand, *etc.*) After this, we adjust the beam so that it lies horizontal without additional weights. We then make marks upon both sides of the beam, at equal distances from the vertical plane passing through the fulcrum. (B) The second is to make the marks on the beam initially. This can easily be done, for instance, by utilizing a graduated ruler as the beam, or by attaching graph paper to a strip of wood, or by marking points equally spaced on a broomstick or bamboo stick, *etc.*, with a pen and ruler. We then fix the nails or hooks above these marks. And finally we place the fulcrum on the plane of symmetry that divides the beam into two equal parts. As we saw before, the fulcrum should not be at the center of the beam. The best place for the fulcrum is along the plane of symmetry, but above the center, in such a way that it is vertically above the CG of the beam, in order to produce a stable equilibrium. After this we must check that the beam remains horizontal when the lever is free to rotate around the fulcrum. If this is not the case, we can attach an appropriate extra weight (a piece of wire, thread or clay) at some point along one of the arms in order to make the beam horizontal.

In Fig. 7.1 we present several kinds of lever, analogous to the balances already built.

Before experimenting with the lever we must test it in order to see if it is calibrated. Let us suppose that it remains horizontal after release without any bodies upon it. We then suspend two equal weights ($P_A = P_B = P$) over two equal arms of the lever ($d_A = d_B = d$). The lever must remain in equilibrium when released from rest horizontally. After this, as we did with the balance, the positions of bodies A and B must be swapped and the lever must remain in equilibrium after release. More-

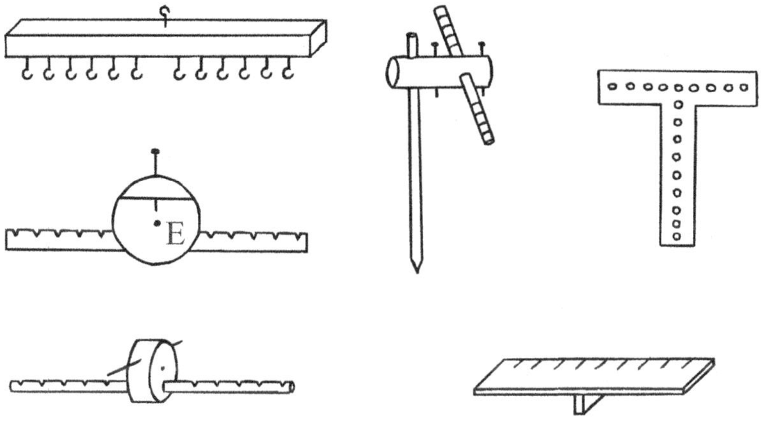

Fig. 7.1. Examples of levers.

over, equilibrium must be maintained for all marks on the lever, that is, for all values of *d*. From now on we will assume that we are working with calibrated levers.

7.2 Experiments with levers and the first law of mechanics

We now begin experimenting with levers.

Experiment
We place a paper clip at the distance of 4 cm from the vertical plane passing through the fulcrum of the lever and another clip of the same weight at a distance of 6 cm from the fulcrum, on the other side of the lever. After the lever is released from rest horizontally, the clip at the larger distance from the fulcrum is observed to fall, while the other rises, as in Fig. 7.2.

The same phenomenon happens for other distances. That is, we place equal weights on arms of different lengths of the lever, $D > d$, releasing the lever from rest horizontally. We again observe that the weight at the larger distance, D, falls, while the other weight rises, as in

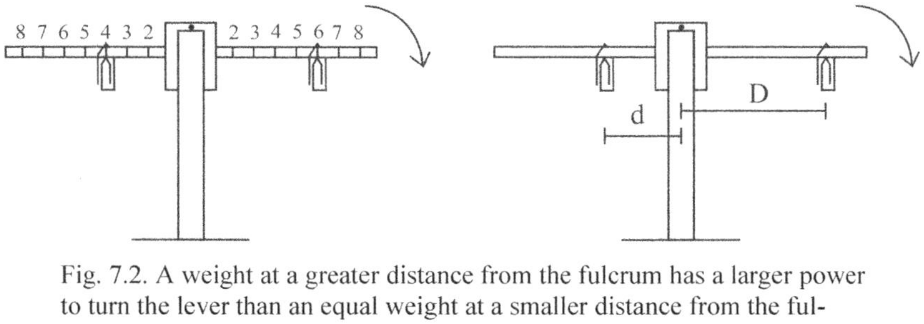

Fig. 7.2. A weight at a greater distance from the fulcrum has a larger power to turn the lever than an equal weight at a smaller distance from the fulcrum.

Fig. 7.2.

This experiment shows that in order to obtain equilibrium, it is not enough to have equal weights on both sides of the fulcrum of a lever. The experiment shows that another relevant factor is the horizontal distance of the weights from the vertical plane passing through the fulcrum. Only experience tells us this; it does not come from theory. That is, experimentally we learn that for the equilibrium of two bodies on a lever the relevant factors are their weights and distances from the fulcrum. On the other hand, other factors do not affect the equilibrium of the lever. Experience teaches that these other irrelevant factors are their colour, shape, texture, chemical composition, volume, *etc.*

This is one of the simplest and most intriguing experiments in mechanics. After all, there are equal weights on both sides of the lever. In spite of this, we observe that the weight at a larger distance from the fulcrum has a greater tendency or power to rotate the lever than the weight at a smaller distance. Although this fact is observed in everyday life, it is still extremely curious.

Experiment

We place 4 paper clips of the same weight at a distance of 6 cm from the fulcrum, equilibrating 4 other identical clips placed at 6 cm from the fulcrum on the other side of the lever. Experience shows that this equilibrium is not disturbed if on one of the sides we place 2 clips at a distance of 4 cm from the fulcrum, and the other 2 clips at a distance of 8 cm from the fulcrum. Equilibrium remains if one of the clips is at a distance of 3 cm from the fulcrum, another at the distance of 5 cm from the fulcrum, while 2 clips remain at a distance of 8 cm from the fulcrum, as in Fig. 7.3.

We can generalize this result as follows. We place N bodies of the same weight at a distance d from the fulcrum, and N other identical bodies at the same distance d from the other side of the fulcrum. The lever remains in equilibrium. Experimentally it is shown that it remains in

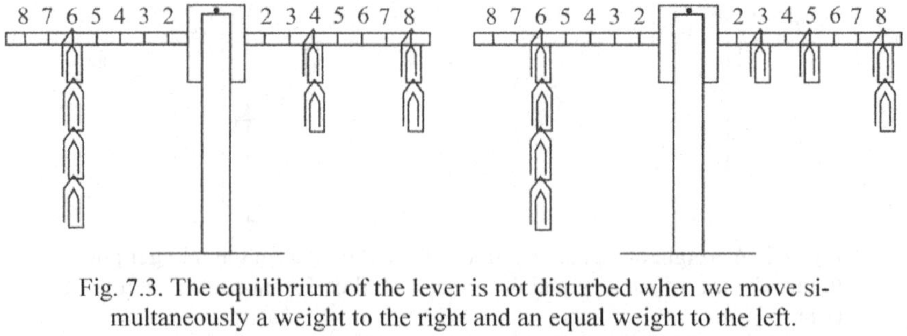

Fig. 7.3. The equilibrium of the lever is not disturbed when we move simultaneously a weight to the right and an equal weight to the left.

equilibrium when we divide one of these groups into two or three parts, with M bodies at the distance d from the fulcrum (M can be equal to zero in the special case), $(N - M)/2$ bodies at a distance $d - x$ from the fulcrum and $(N - M)/2$ of these bodies at a distance $d + x$ from the fulcrum. On the other hand, equilibrium will not occur if we place $(N - M)/2$ bodies at a distance $d - x_1$ from the fulcrum and $(N - M)/2$ of these bodies at a distance $d + x_2$ from the fulcrum, if x_1 is different from x_2. Equilibrium will remain in the first case if we can divide one or more of the first groups of $(N - M)/2$ bodies into two or three sub-groups, by placing Q of them at the distance $d - x$ from the fulcrum, while $((N - M)/2 - Q)/2$ are placed at a distance $(d - x) - y$ from the fulcrum and $((N - M)/2 - Q)/2$ are placed at a distance $(d - x) + y$ from the fulcrum. And so on. In the previous example we had $N = 4$, $M = Q = 0$, $d = 6$ cm, $x = 2$ cm and $y = 1$ cm.

This experiment is not trivial. It shows that a weight P placed at a distance d from the fulcrum is equivalent to a weight $P/2$ placed at a distance $d - x$ from the fulcrum, together with another weight $P/2$ at a distance $d + x$ from the fulcrum. That is, these two weights $P/2$ on one side of the fulcrum, at distances $d + x$ and $d - x$ from it, equilibrate a weight P on the other side of the fulcrum at a distance d from it. This experiment indicates that, as regards the rotation of the lever, the weights act independently of one another, following the principle of superposition, with a linear influence of their distances from the fulcrum. If the influences of their distances to the fulcrum were not linear but followed another law (quadratic, cubic, inverse of the distance, inverse square, sinusoidal, logarithmic, *etc.*), then the equivalence already observed would no longer hold. Once more, this comes from experiment; no logical argument obliges nature to behave like this.

We now analyze the equilibrium of a lever with different weights on its two arms.

Experiment

We take 5 paper clips of the same weight. We place 2 of these clips at a distance of 6 cm from the vertical plane passing through the fulcrum. We place the other 3 clips at a distance of 6 cm from the other side of the vertical plane passing through the fulcrum, releasing the lever from rest horizontally. We observe that it turns around the fulcrum, with the 3 clips falling downward and the 2 clips rising upward, as in Fig. 7.4.

This experiment can be generalized to other cases. That is, suppose we have N bodies of equal weight P at a distance d from one side of the vertical plane passing through the fulcrum of a lever, and M other bodies

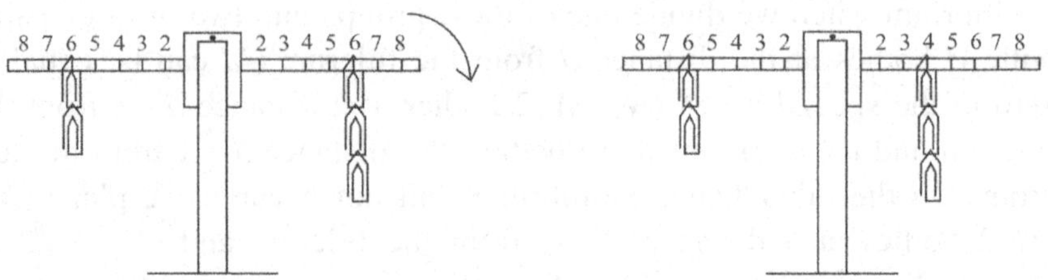

Fig. 7.4. Equilibrium of different weights.

of equal weight P at the same distance on the other side of the fulcrum, with $M > N$. If we release the lever from rest horizontally, it turns around the fulcrum, with the set of M bodies falling downward and the set of N bodies rising upward. By the definitions we introduced earlier, we say that the set of M bodies is heavier than the set of N bodies.

Now comes one of the most important experiments of all.

Experiment

We consider 5 clips of the same weight. We place 2 of these clips at the same distance of 6 cm from the vertical plane passing through the fulcrum. We want to find the distance from the other side of the fulcrum at which we should place the 3 other clips together in order to place the lever in equilibrium (that is, at rest horizontally after released). Experiment shows that this only happens when they are at a distance of 4 cm from the vertical plane passing through the fulcrum, as in Fig. 7.4.

If we place the 2 clips at the same distance d_A from the vertical plane passing through the fulcrum, we observe that the balance only remains in equilibrium after release if the other 3 clips acting together on the other side of the fulcrum are located at a distance d_B as given by the following table.

d_A (cm)	2	3	4	5	6	7	8
d_B (cm)	4/3	2	8/3	10/3	4	14/3	16/3

7.2.1 First part of the law of the lever

The result of this experiment is also verified in other cases. We place N_A bodies of the same weight P together on the arm of length d_A of a lever. Their total weight is given by $P_A \equiv N_A P$. We place N_B other bodies of the same weight P together on the other side of the lever at a distance d_B from the fulcrum, releasing the lever from rest horizontally. Experiment shows that it only remains in equilibrium if

$$\frac{d_B}{d_A} = \frac{P_A}{P_B} = \frac{N_A}{N_B} .$$

This is the initial part of the first law of mechanics. Sometimes it is called the law of the lever. Archimedes expressed it as follows in Proposition 6 of the first part of his work *On the Equilibrium of Planes*: "Commensurable magnitudes are in equilibrium at distances reciprocally proportional to the weights."[1]

By "magnitudes" we understand that Archimedes was referring to physical bodies. The idea behind commensurable magnitudes is measurement by comparison. That is, to measure two or more magnitudes with the same unit or standard of measure. If the ratio between the weights of two bodies is equal to the ratio of two integers, like $5/3$, we say that the weights are commensurable. The reason for this is that in this case it is possible to find a common measure, so that the weight of each one of these bodies is an integer multiple of this common measure. The body A might measure, for instance, 5 N (five Newtons), while body B might measure 3 N. In this example the unit of measure might be one Newton = 1 N, or 0.01 N, or ... The same ratio of $5/3$ would occur if the weight of A was 20 N and that of B was 12 N. Also in this case the unit of measure might be 1 N, or 0.01 N, *etc.* If the ratio of the weights of two bodies is equal to the ratio between an irrational number (like the square root of 2, or like the number π) and an integer, we then say that these two weights are incommensurable. In this case it is not possible to find a common measure such that the weight of each one of these bodies would be an integer multiple of this common measure. The length of the diagonal of a square, for instance, is incommensurable with the length of a side of the square. The same holds for the ratio between the area of a circle of unit diameter and the area of a square of unit sides.

Archimedes generalized this result for incommensurable magnitudes in Prop. 7 of this work: "However, even if the magnitudes are incommensurable, they will be in equilibrium at distances reciprocally proportional to the magnitudes."[2]

In his English translation of Archimedes's work, Heath combined these two propositions in a single one, namely: "Propositions 6, 7. Two magnitudes, whether commensurable [Prop. 6] or incommensurable [Prop. 7], balance at distances reciprocally proportional to the magnitudes."[3]

As this is one of the most important laws of classical mechanics, it is worth while calling attention to some experimental errors which prevent the verification of this result.

---

1 E.J. Dijksterhuis, *Archimedes* (Princeton: Princeton University Press, 1987), p. 289.
2 E.J. Dijksterhuis, *Archimedes*, p. 305.
3 T.L. Heath, ed., *The Works of Archimedes* (New York: Dover, 2002), p. 192.

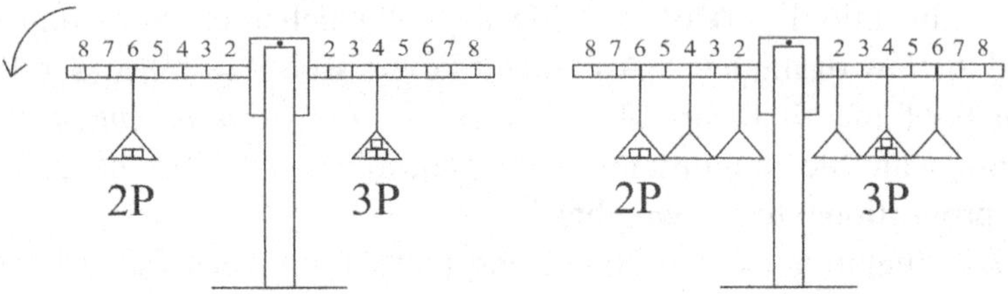

Fig. 7.5. How to observe the law of the lever correctly.

Let us suppose that the lever remains initially at rest horizontally without the scale pans and also without bodies A and B. Let us suppose that both scale pans and their threads have the same weight. The most frequent mistake is to place weight P_A on one of the scale pans at a distance d_A from the fulcrum, and another weight P_B on the other scale pan at a distance d_B on the other side of the fulcrum. In this case, the lever does not remain in equilibrium after release even if $d_B/d_A = P_A/P_B$. What happens is that the larger arm moves downward (assuming a lever with high sensitivity, with negligible friction, totally free to rotate around the fulcrum), as in Fig. 7.5.

The explanation of this phenomenon is related to the law of the lever itself. Although bodies A and B balance one another when placed at distances inversely proportional to their weights, the same does not hold for the two equal scale pans. Here we have two scale pans of the same weight placed at different distances from the fulcrum. By the previous experiments we know that they do not balance one another. Instead of this, the larger arm moves downward. In order to prevent this common mistake, we did not employ any scale pans in the experiments with levers performed thus far. Instead, we suspended the bodies directly from the beam. But it is possible to utilize scale pans in a lever, provided they are equal in number on both sides, with each pair of equal scale pans placed at the same distance from the fulcrum. For example, we can have 6 equal scale pans, three of them placed at distances of 2 cm, 4 cm and 6 cm from one side of the fulcrum, and the other three placed at the same distances on the other side. In this case the lever remains in equilibrium even after A and B are placed on the scale pans, provided $d_B/d_A = P_A/P_B$, as in Fig. 7.5.

Another common mistake that is made even without scale pans is as follows. Suppose that a lever remains in equilibrium when the vertical plane passing through the fulcrum divides the homogeneous beam into two equal parts. We now place two bodies of different weights at the extremities of the beam and change the location of the fulcrum in such a way that $d_B/d_A = P_A/P_B$. The lever does not remain in equilibrium

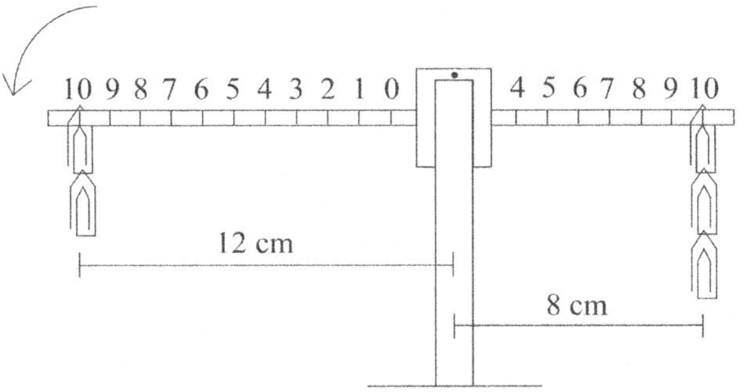

Fig. 7.6. This is not the correct way to observe the law of the lever.

in this case. Instead of this, the side with longer arm falls downward, as in Fig. 7.6.

Once more, the explanation for this behaviour is related to the law of the lever itself. Let us suppose that there are no scale pans and that the bodies A and B are suspended directly from the beam following the previous relation. Therefore they balance the fulcrum in its new position because they satisfy the relation $d_B/d_A = P_A/P_B$. But the beam itself is not in equilibrium for the new position of the fulcrum. When we changed the position of the fulcrum in relation to the center of the beam, the beam became unbalanced, regardless of the positions of bodies A and B. The longer arm of the homogeneous beam tends to fall downward, as it is heavier than the other side, which moves upward. Even placing bodies A and B on the beam satisfying the previous relation does not balance the beam. In order to avoid this mistake, the correct procedure is to balance the beam without bodies A and B, adjusting the fulcrum over the CG of the beam in such a way that it remains horizontally at rest relative to the ground. After this, without changing the position of the fulcrum in relation to the beam, we can place bodies A and B. In this case it will be seen that they will keep the beam in balance provided that $d_B/d_A = P_A/P_B$.

These two mistakes are related to the fact that the scale pans and the beam itself are material bodies with weight. Therefore, they may also influence the equilibrium of the lever. This aspect cannot be neglected when we work with sensitive levers and wish to identify precisely which quantitative factors determine the equilibrium of bodies.

Suppose now that P_A/P_B is different from d_B/d_A. In this case there will be no equilibrium when the lever is released from rest. One of the bodies will move downward, and the other, upward. The experimental results already presented can be summarized by saying that if

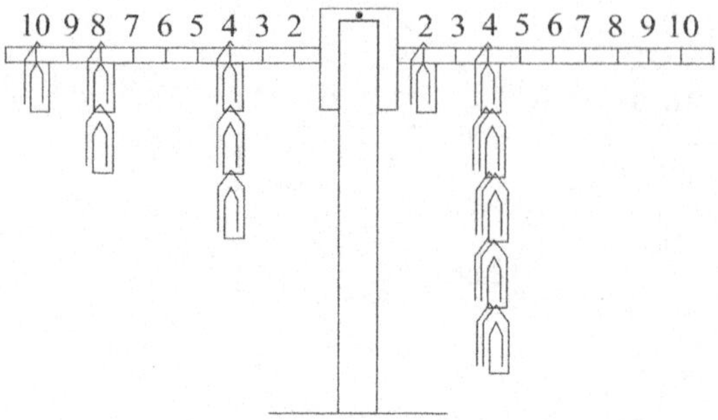

Fig. 7.7. Principle of superposition.

$(P_A/P_B)(d_A/d_B) > 1$, then A will fall and B will rise. If $(P_A/P_B)(d_A/d_B) < 1$, then A will rise and B will fall.

7.2.2 Second part of the law of the lever

Experiment

We take 16 paper clips of the same weight. On one side of the lever we place 1 clip at a distance of 10 cm from the fulcrum, 2 clips at 8 cm from the fulcrum, and 3 clips at 4 cm from the fulcrum. On the other side of the lever we put 1 clip at 2 cm from the fulcrum and 9 clips at 4 cm from the fulcrum. It is observed that the lever remains in equilibrium, as in Fig. 7.7.

This experiment shows that as regards rotation of the lever, the weights act independently of one another, proportionately to their distances to the fulcrum. That is, the rotation effects due to the weights follow the law of addition. This is expressed in physics by saying that the law of the lever follows the principle of superposition.

The result of this specific experiment is also true in other cases, and can be generalized as follows. We place N weights $P_1, P_2, ..., P_N$ on one side of the lever, at distances $d_1, d_2, ..., d_N$, respectively, from the vertical plane passing through the fulcrum. We place M other weights $P_{N+1}, P_{N+2}, ..., P_{N+M}$ on the other side of the fulcrum, at distances $d_{N+1}, d_{N+2}, ..., d_{N+M}$, respectively, from the vertical plane passing through the fulcrum. We observe that the system only remains in equilibrium after release from rest horizontally if

$$\sum_{i=1}^{N} \frac{P_i}{P_0} \frac{d_i}{d_0} = \sum_{i=N+1}^{N+M} \frac{P_i}{P_0} \frac{d_i}{d_0} .$$

Here P_0 and d_0 are a weight and a distance, chosen arbitrarily. We can have, for instance, $P_0 = P_1$ and $d_0 = d_1$. Or we can choose $P_0 = P_2$ and $d_0 = d_2$, and so on.

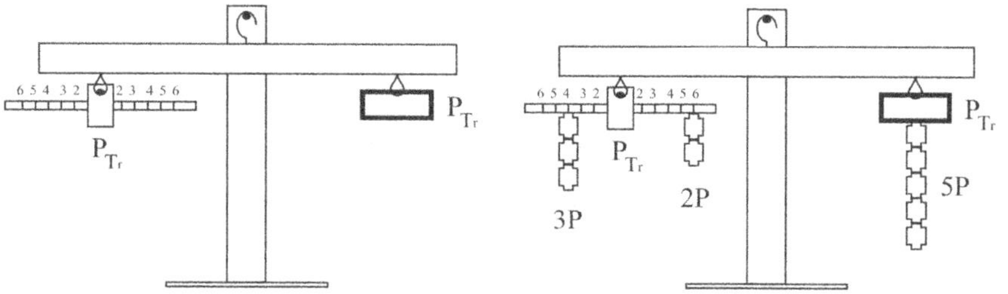

Fig. 7.8. Levers in equilibrium.

This is the final part of the first law of mechanics, *i.e.*, the law of the lever combined with the principle of superposition.

To illustrate, in the previous example let P_0 be the weight of a clip and $d_0 = 1$ cm. The left side the previous relation yields: $1\times 10+2\times 8+3\times 4=38$. On the right side we have: $1\times 2+9\times 4=38$. This demonstrates the state of equilibrium.

Experiment

We suspend a lever by the fulcrum on one of the sides of a balance with equal arms, in such a way that the lever remains horizontal without extra weights. On the other side of the balance we suspend a weight P_{tr} equal to the weight of the lever, such that the balance remains in equilibrium horizontally. We then take 10 paper clips of the same weight. We place 3 of them on one arm of the lever at a distance of 4 cm from the fulcrum, and 2 of them on the other side of the lever at a distance of 6 cm from the fulcrum. We then try to find how many clips of the same weight we need to suspend on the other side of the balance in order to keep it in equilibrium. Experimentally it is found that this only happens by hanging 5 clips, as in Fig. 7.8.

This and other analogous experiments show that the fulcrum of a balance in equilibrium with weights P_A and P_B at the distances d_A and d_B, respectively, on opposite sides of the fulcrum, in such a way that $P_A/P_B = d_B/d_A$, supports a total weight of $P_{tr} + P_A + P_B$. Here P_{tr} is the weight of the lever. We can then see that there are four forces acting upon the beam of a lever in equilibrium: (A) the downward weight of the beam acting as if it were concentrated at the CG of the beam; (B) the downward weight of body A acting at a distance d_A from the fulcrum; (C) the downward weight of body B acting at a distance d_B from the other side of the fulcrum; and (D) the normal upward force N acting along the fulcrum, as in Fig. 7.9.

The weights of the beam and of bodies A and B are due to their gravitational interactions with the Earth. The normal upward force is ex-

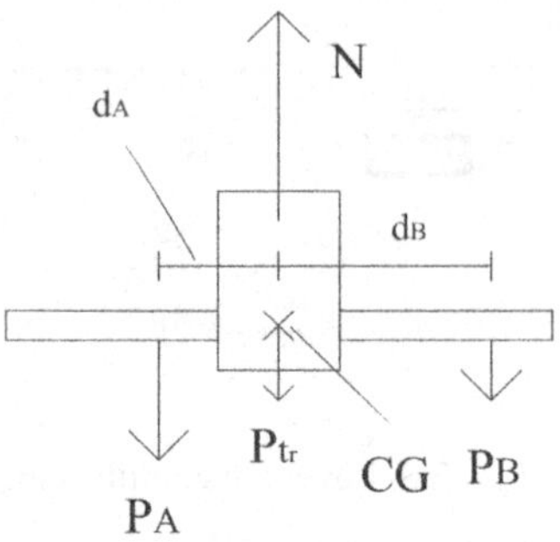

Fig. 7.9. Conditions of equilibrium for a lever.

erted by the support upon the beam and arises due to the tension or compression of the support. The support will be stretched or under mechanical tension when it is a hook (or a thread, or a spring) attached to a rigid support at its upper end, holding the fulcrum of the lever at its lower end, as in the previous experiment. The support will be compressed when it is a rigid stand or a spring placed below the fulcrum, as in the majority of the situations considered up to now. We then see that there are two conditions for a balanced lever:

$$N = P_{Tr} + P_A + P_B ,$$

$$\frac{P_A}{P_B} = \frac{d_B}{d_A} .$$

The latter relation needs to be generalized if the fulcrum is not along the same vertical plane passing through the CG of the beam. Let us suppose that the CG of the lever is along the same side of the vertical plane passing through the fulcrum as the body B, at a distance d_{Tr} from this plane, as in Fig. 7.10.

In this case the conditions for balancing the lever are given by:

$$N = P_{Tr} + P_A + P_B ,$$

$$\frac{P_A}{P_0}\frac{d_A}{d_0} = \frac{P_{Tr}}{P_0}\frac{d_{Tr}}{d_0} + \frac{P_B}{P_0}\frac{d_B}{d_0} .$$

Once more, P_0 and d_0 are a weight and a distance, chosen arbitrarily.

If $d_{Tr} = 0$, or if we can neglect the weight of the lever in comparison with the weights of bodies A and B, we return to the previous case.

When we have several bodies acting on the lever we can utilize the principle of superposition given earlier in order to establish the equilibrium conditions.

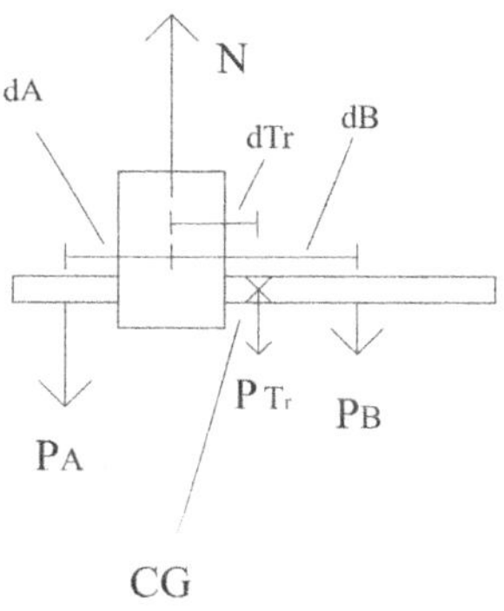

Fig. 7.10. Conditions of equilibrium for an asymmetrical lever.

Complement to the law of the lever: The downward force exerted by the fulcrum upon the support when the lever is in equilibrium is given by the sum of the weights of the suspended bodies, plus the weight of the lever (that is, of its beam, threads, and scale pans).

7.3 Types of levers

We saw earlier how to utilize a balance of equal arms in order to compare the force due to weight with other forces of any nature (contact forces, elastic forces, electromagnetic forces, *etc.*) That is, a force F acting on one side of a balance and equilibrating a weight P on the other side is defined as equal to this weight. This operational definition of force, together with the law of the lever, is related to the utilization of the lever as a simple machine. The law of the lever shows that a small weight can equilibrate a large weight provided it is at a greater distance from the fulcrum than the large weight. A simple machine is a device that can multiply the intensity of a force in order to do work.

In this section we will neglect the weight of the lever as compared with the other forces acting upon it.

The law of the lever states that a weight P_A located at distance d_A from the vertical plane passing through the fulcrum equilibrates another weight P_B at distance d_B from the vertical plane passing through the fulcrum if $P_A/P_B = d_B/d_A$. When we utilize a lever as a simple machine, it is more convenient to talk of forces than of weights, as the forces acting upon the lever do not need to be gravitational in origin. Let F_A be the force exerted upon the machine by the operator (a man, an animal, or a mechanical device) and F_R be the resistive force. That is, F_R is the force exerted by the machine upon the load (weight to be raised or pushed, body to be compressed or stretched, figure to be cut out, *etc.*) To sim-

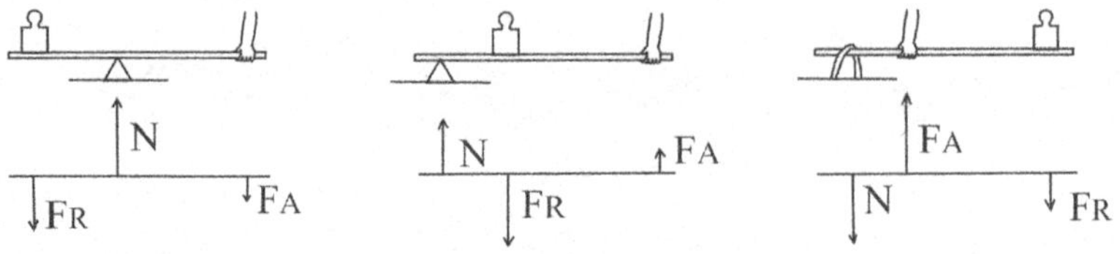

Fig. 7.11. Types of lever.

plify the analysis we will suppose that the points of application of F_A and F_R are aligned with the fulcrum of the lever, and that these two forces act orthogonally to this straight line. The arms of the lever (that is, the distances between the points of application of these forces and the fulcrum) will be represented by d_A and d_R, respectively. The equilibrium of the lever is then given by $F_A/F_R = d_R/d_A$.

We can see that there are three main elements to a lever working as a simple machine: the applied force (the effort), the resistive force (the load) and the fulcrum, which remains at rest relative to the Earth. Depending upon the position of the fulcrum in relation to the applied and resistive forces, there will be three basic kinds of lever, as in Fig. 7.11.

A. Lever of the first class, with the fulcrum between the load and the effort.
B. Lever of the second class, with the load between the effort and the fulcrum.
C. Lever of the third class, with the effort between the fulcrum and the load.

Up to now we have worked only with levers of the first kind. Examples of this kind of lever are the equal arm balance, Roman balance, seesaw, the crowbar, salad tongs, pair of pliers, scissors, handle of a water pump, hammer used to pull a nail out of wood, *etc.*

Examples of levers of the second class are the wheelbarrow, bottle opener, door, pair of nutcrackers, punching machine, paper cutter, *etc.*

Examples of levers of the third class are a pair of tweezers, broom, pair of barbecue tongs, forceps, jaws, the human forearm, fishing pole, stapler, *etc.*

Chapter 8

Mathematical Definition of Center of Gravity

The law of the lever and the principle of superposition allow a mathematical definition of the center of gravity of a body or of a system of bodies. We saw earlier that the condition of equilibrium of any body suspended by a horizontal axis (the fulcrum or PS) is that this axis and the CG of the body should be along a vertical. The equilibrium will be stable (unstable) if any perturbation in the position of the body raises (lowers) the CG from its previous position.

We now consider a lever in stable equilibrium with its beam resting horizontally, without other bodies suspended on it. We imagine a homogeneous beam in such a way that the vertical plane passing through the fulcrum divides it into two equal halves. The CG of the beam is vertically below the fulcrum. We have seen that this equilibrium is not disturbed if two bodies A and B of weights P_A and P_B, respectively, are suspended on opposite sides of the fulcrum, provided that $d_B/d_A = P_A/P_B$. Here d_A and d_B are the horizontal distances between the points of suspension of A and B, respectively, and the vertical plane passing through the fulcrum. This means that the CG of these two bodies is also along the vertical plane passing through the fulcrum. If the ratio d_B/d_A is different from P_A/P_B, the beam does not remain in equilibrium after being released.

In order to find an algebraic expression yielding the location of the CG of the bodies A and B we can imagine a horizontal axis x along the beam. The origin $x = 0$ can be chosen at any point, arbitrarily. Let us suppose that the ends of the beam of length L are located at x_E and $x_D = x_E + L$. Let x_A and x_B be the points of suspension of bodies A and B along the x axis, respectively, as in Fig. 8.1. Moreover, let us assume

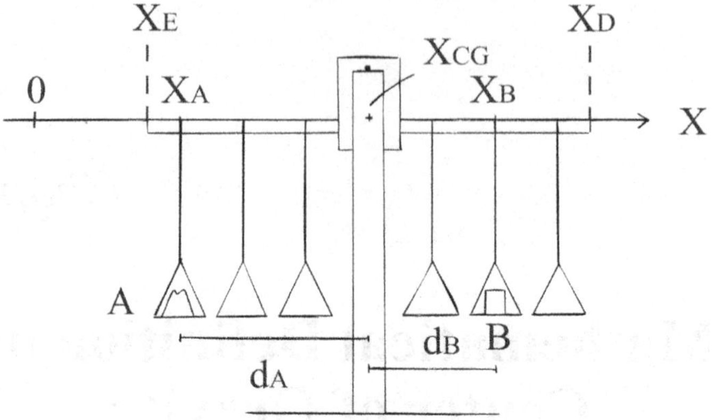

Fig. 8.1. Finding an algebraic expression for the center of gravity.

that the lever continues in equilibrium after being released from rest horizontally with A and B acting upon these points.

The CG of this system must be along the vertical plane passing through the fulcrum, in such a way that $d_B/d_A = P_A/P_B$. Let x_{CG} be the location of the CG of bodies A and B along the x axis. From Fig. 8.1 we have $d_A = x_{CG} - x_A$ and $d_B = x_B - x_{CG}$. From the law of the lever we can then define, mathematically, the position x_{CG} of the center of gravity of this system of two bodies along the x axis as given by

$$\frac{x_B - x_{CG}}{x_{CG} - x_A} \equiv \frac{P_A}{P_B} .$$

That is,

$$x_{CG} = \frac{P_A}{P_T} x_A + \frac{P_B}{P_T} x_B ,$$

where $P_T \equiv P_A + P_B$ is the total weight of the two bodies.

This theoretical definition of x_{CG} is made in such a way that it coincides with the previous experimental results on the CG of rigid bodies. In other words, in equilibrium the CG of the system of two bodies stays along the vertical plane passing through the fulcrum of the lever. If $P_A = P_B$, we can see from this expression that x_{CG} will be at the midpoint between x_A and x_B. On the other hand, the larger (smaller) the value of P_A/P_B, the closer (farther) x_{CG} will be from body A.

From now on we will use the approximation of particles or point bodies. We consider bodies A and B as particles when the greatest dimensions of either (their diameters, or the greatest distance between any material points belonging to each one of these bodies) are much smaller than the distance between A and B. In this case we can treat the bodies

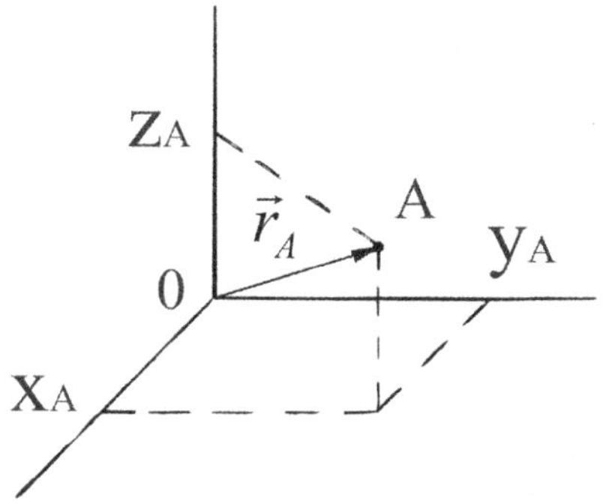

Fig. 8.2. Finding the CG with vector notation.

as being concentrated in small regions as compared to the distance between them, as if they were concentrated into mathematical points.

Let us now imagine a rigid system of orthogonal axes xyz with origin O at $x = y = z = 0$. The spatial location of body A will be represented by (x_A, y_A, z_A), and that of body B by (x_B, y_B, z_B). In this way we can generalize the previous relation for the CG of the system for the y and z axes. We then define the y and z coordinates of the CG, y_{CG} and z_{CG}, respectively, by the relations

$$y_{CG} = \frac{P_A}{P_T} y_A + \frac{P_B}{P_T} y_B ,$$

$$z_{CG} = \frac{P_A}{P_T} z_A + \frac{P_B}{P_T} z_B .$$

In this way we can also utilize vector notation. We call $\vec{r}_A = (x_A, y_A, z_A)$ the position vector of body A, as in Fig. 8.2, and $\vec{r}_B = (x_B, y_B, z_B)$ the position vector of body B.

The position vector of the CG, $\vec{r}_{CG}$, is defined by:

$$\vec{r}_{CG} \equiv \frac{P_A}{P_T} \vec{r}_A + \frac{P_B}{P_T} \vec{r}_B .$$

By the principle of superposition these relations can be extended to a set of N particles. Let P_i be the weight of body i located at (x_i, y_i, z_i), with $i = 1, 2, ..., N$. The x component of the CG of this system of particles is defined by (with P_T being the total weight of this system of particles)

$$x_{CG} \equiv \sum_{i=1}^{N} \frac{P_i}{\sum_{j=1}^{N} P_j} x_i \equiv \sum_{i=1}^{N} \frac{P_i}{P_T} x_i .$$

Analogous expressions are defined for the y and z components of the CG.

The position vector of the CG of this system of point particles is defined by

$$\vec{r}_{CG} \equiv \sum_{i=1}^{N} \frac{P_i}{P_T} \vec{r}_i \, .$$

This is the modern mathematical definition of the CG of a system of particles. It makes a theoretical calculation of the CG possible, if the locations of particles and their weights are known.

If we have continuous distributions of matter, as in the case of one-, two- and three-dimensional bodies, the procedure is the same. In the first place we replace the summation by line, surface or volume integrals. And instead of the weight P_i of particle i we utilize an infinitesimal element of weight, dP, located at $\vec{r} = (x, y, z)$. This element of weight dP represents the weight contained in an infinitesimal element of length, area or volume. The total weight is given by $P_T = \iiint dP$. In this case the position vector of the CG can be defined by

$$\vec{r}_{CG} \equiv \iiint \vec{r} \frac{dP}{P_T} \, .$$

These volume integrals should be performed over the whole space occupied by the body. If we have matter distributed continuously along a line or surface, we replace these volume integrals by line or surface integrals, respectively.

If we have combinations of discrete and continuous distributions of matter, we only need to add the corresponding expressions in order to obtain the CG of the system as a whole, because the CG follows the principle of superposition.

We will not go into mathematical details here, nor will we calculate the location of the CG for any distribution of matter, as this is not the goal of this book.

We can summarize the modern mathematical definition of the CG as follows.

Mathematical definition CG9:

$$\vec{r}_{CG} \equiv \sum_{i=1}^{N} \frac{P_i}{P_T} \vec{r}_i \, ,$$

or

$$\vec{r}_{CG} \equiv \iiint \vec{r} \frac{dP}{P_T} \, .$$

These are the theoretical definitions in current use to calculate the CG of discrete and continuous distributions of matter, when the weights and locations of the bodies are known.

An important theorem which simplifies the location of the center of gravity states the following, adapted by Symon.[1] If a body is composed of two or more parts whose centers of gravity are known, then the center of gravity of the composite body can be calculated by regarding its component parts as single particles located at their respective centers of gravity. It was essentially this theorem that Archimedes utilized to calculate the centers of gravity of many geometric homogeneous figures. He considered it a postulate, as we will see later on.

A proof of this theorem, beginning with definition CG9, can be given as follows. Let a body be composed of N parts of weights P_1, ..., P_N. Let any part P_k be composed of N_k parts of weights P_{k1}, ..., P_{kN_k}, whose centers of gravity are located at the points $\vec{r}_{k1}$, ..., $\vec{r}_{kN_k}$. Then the center of gravity of the part P_k is located at the point

$$\vec{r}_k \equiv \sum_{\ell=1}^{N_k} \frac{P_{k\ell}}{P_k} \vec{r}_{k\ell} ,$$

where

$$P_k \equiv \sum_{\ell=1}^{N_k} P_{k\ell} .$$

The center of gravity of the entire body is located at the point

$$\vec{r}_{CG} \equiv \sum_{k=1}^{N} \sum_{\ell=1}^{N_k} \frac{P_{k\ell}}{P_T} \vec{r}_{k\ell} ,$$

where

$$P_T \equiv \sum_{k=1}^{N} \sum_{\ell=1}^{N_k} P_{k\ell} .$$

This means that the center of gravity of the entire body can be written as

$$\vec{r}_{CG} \equiv \sum_{k=1}^{N} \sum_{\ell=1}^{N_k} \frac{P_{k\ell}}{P_T} \vec{r}_{k\ell} = \sum_{k=1}^{N} \frac{P_k}{P_T} \left(\sum_{\ell=1}^{N_k} \frac{P_{k\ell}}{P_k} \vec{r}_{k\ell} \right) = \sum_{k=1}^{N} \frac{P_k}{P_T} \vec{r}_k ,$$

where the total weight can also be written as

$$P_T \equiv \sum_{k=1}^{N} \sum_{\ell=1}^{N_k} P_{k\ell} = \sum_{k=1}^{N} P_k .$$

These two last equations embody the mathematical statement of the theorem to be proved.

[1] K.R. Symon, *Mechanics* (Reading: Addison-Wesley, 1971), third edition, p. 221.

Chapter 9

Explanations of and Deductions from the Law of the Lever

9.1 Law of the lever as an experimental result

What we have seen so far constitutes the most important aspects of statics. We can summarize the subject as follows:

Definitions: We say that an equal arm balance and a lever are in equilibrium when their arms remain at rest horizontally, with the beam free to rotate around the fulcrum. Two bodies A and B have the same weight P if they keep this balance in equilibrium after being placed on its separate scale pans and released from rest. The body which equilibrates N other bodies of the same weight P on an equal arm balance has N times the weight P.

Experimental results: Two bodies of weights P_A and P_B equilibrate one another on opposite sides of a horizontal lever which has the CG of the beam along the vertical plane passing through the fulcrum, if $P_A/P_B = d_B/d_A$. Here d_A and d_B are the horizontal distances between the points of suspension bodies A and B, respectively, and the vertical plane passing through the fulcrum. If we have N bodies acting upon one side of the lever and M bodies acting on the other side, the equilibrium can be obtained by the principle of superposition, assuming that the weights act independently of one another in such a way that we can add their individual contributions. This means that there will be equilibrium if the following relation is valid:

$$\sum_{i=1}^{N} \frac{P_i}{P_0}\frac{d_i}{d_0} = \sum_{i=N+1}^{N+M} \frac{P_i}{P_0}\frac{d_i}{d_0} .$$

Here P_0 and d_0 are a weight and a distance, chosen arbitrarily. We can derive an interesting result from this latter condition of equilibrium. Let us suppose that on one of the sides of a lever in equilibrium we have two equal weights $P_1 = P_2 = P$ acting at distances $d_1 = d - x$ and $d_2 = d + x$ from the fulcrum. It is easy to see that

$$\frac{P_1}{P_0}\frac{d_1}{d_0} + \frac{P_2}{P_0}\frac{d_2}{d_0} = \frac{P}{P_0}\frac{d-x}{d_0} + \frac{P}{P_0}\frac{d+x}{d_0} = \frac{2P}{P_0}\frac{d}{d_0} = \frac{P}{P_0}\frac{2d}{d_0} .$$

These two weights P_1 and P_2 are equivalent to a single weight $P_3 = 2P$ acting at a distance $d_3 = d$ from the fulcrum, or to a single weight $P_4 = P$ acting at a distance $d_4 = 2d$ from the fulcrum. The equivalence here refers to the tendency to rotate the lever. In other words, if P_1 and P_2 keep the lever in equilibrium, then $P_3 = 2P$ acting at $d_3 = d$ will also keep it in equilibrium. The same is valid for $P_4 = P$ acting at $d_4 = 2d$. Later on we will see that we can invert this situation. We can begin with the equivalence of P_3 to the set P_1 and P_2, and arrive at the law of the lever.

With the previous mathematical law of the lever we can explain the experimental result that in equilibrium, the CG of a rigid body is along the vertical line passing through the PS. Because the mathematical expression of the CG, *i.e.*, CG9, was defined according to the law of the lever, this result follows automatically.

It is possible to utilize the law of the lever to deduce more complex situations. That is, we do not need to explain the law of the lever; we can simply accept it as an empirical fact of nature. We postulate the mathematical relation by saying that it agrees with the experimental data. We then utilize this law in order to explain the mechanism behind many types of toys and simple machines (such as the equilibrist and toys we saw earlier, or levers of the first, second and third classes). This is the simplest procedure, and there are no problems in assuming this point of view.

Another alternative is to try to derive the law of the lever experimentally or theoretically. For this, we need to begin with other experimental results, or we need to create other concepts and theoretical postulates. One motivation for following this route is that we want to find a simpler way to arrive at the law of the lever. An opposite motivation might be to begin with something more complex or more abstract than the law of the lever itself, in order to arrive not only at this law but also at other relevant results. For instance, it may be possible to utilize these new concepts and postulates to also arrive at results which are independent of the law of the lever, like the law of the inclined plane. Another reason to follow this new procedure is that we can utilize these new concepts and postulates in order to arrive at other laws and physical results which are valid not only in conditions of equilibrium but also, for instance, when the bodies are in motion in relation to the Earth. This might be the case, for instance, if we were studying the more general laws that govern the rotation and acceleration of rigid bodies relative to the Earth.

Whenever we follow this alternative procedure, it should be kept in mind that we cannot explain everything. We can postulate the law of the lever (L) without explaining it and derive consequences (C_1), (C_2), *etc.*, from it. Or, alternatively, we can postulate some other law (P) without explaining it and derive the results (L), (C_1), *etc.*, from it. The crux is that in all procedures we always need to begin with some axiom or postulate (which has no explanation) in order to explain other things from it. The only justification of the basic axioms or postulates may be that they agree with experimental data or that they lead to verifiable experimental data.

In the next Sections we will see different ways to derive the law of the lever from other experimental results, or from other theoretical postulates. There are still other ways to derive this law which will not be considered here.

9.2 Deriving the law of the lever from the torque concept

Earlier we saw the first condition of equilibrium for a body to remain at rest relative to the Earth, in the presence of gravity. This condition is that the downward weight P acting upon the body must be counterbalanced by another upward force N, of the same magnitude as the weight. This prevents the motion of the body as a whole relative to the ground, if it begins from rest. In the case of the balance or lever, we have a horizontal axis fixed relative to the ground, its fulcrum. Therefore, the weight of the bodies placed upon the beam, together with the weight of the beam itself, must be counterbalanced by an upward normal force N acting at the fulcrum, exerted by the support of the lever. Nevertheless, the balance or the lever can turn around the fulcrum.

We have seen that the concept of weight is not sufficient for the equilibrium of the lever. After all, two bodies of the same weight but placed on opposite sides of the fulcrum, at different distances from it, disturb the equilibrium of the lever. In this case the body acting at a larger distance from the fulcrum will fall toward the ground, with the other weight moving away from it, even though the fulcrum remains at rest relative to the ground. This shows that equal weights acting at different distances from the fulcrum tend to turn the lever.

Due to this fact we conclude that we need another concept, beyond the net force acting upon a rigid body, in order to establish the conditions of equilibrium of this body. This rigid body could be, for instance, the beam of a lever. We can utilize the lever in order to define this new concept related to the rotation of a rigid body around a horizontal axis which is fixed relative to the ground. Let us suppose the simplest case in which the fulcrum of the lever (that is, the horizontal axis around which it can turn) is vertically above the CG of the lever. We then suppose two new forces F_A and F_B acting in the same sense, vertically downwards, at horizontal distances d_A and d_B, respectively, from the vertical plane passing through the fulcrum. The experimental law of the lever informs us that if this lever is released from rest horizontally, being free to rotate around the fulcrum, it will remain at rest under the action of these two forces only if $F_A/F_B = d_B/d_A$.

We then define what causes the rotation of a rigid body around a horizontal axis which is fixed in relation to the ground as the "torque or moment of a force." We will represent this torque or moment by the letter T. The experimental law of the lever allows us to define the quantitative ratio T_A/T_B between the magnitudes of the torques exerted by the two forces F_A and F_B already mentioned as

$$\frac{T_A}{T_B} \equiv \frac{F_A}{F_B}\frac{d_A}{d_B} .$$

This definition was suggested by an experimental result. But now that we have this definition, we can reverse the argument. The usual procedure is to postulate that the lever will remain in equilibrium if $T_A = T_B$. This postulate and the previous definition of the ratio of the magnitudes of two torques leads to the law of the lever, namely, $(F_A/F_B)(d_A/d_B) = 1$. If $T_A/T_B > 1$ and the lever is released from rest horizontally, we postulate that body A will move towards the ground and body B will move away from it. If $T_A/T_B < 1$ and the lever is released from rest horizontally, we postulate that body A will move away from the ground and body B will move towards it.

It may seem that we do not gain anything with this theoretical deduction. After all, we are defining the ratio of torques according to the law of the lever. And in the end we are arriving at the law of the lever itself, by postulating that in equilibrium the torques acting on both sides of the lever have the same magnitude. But as already mentioned, this procedure may have some advantages if we utilize this concept not only

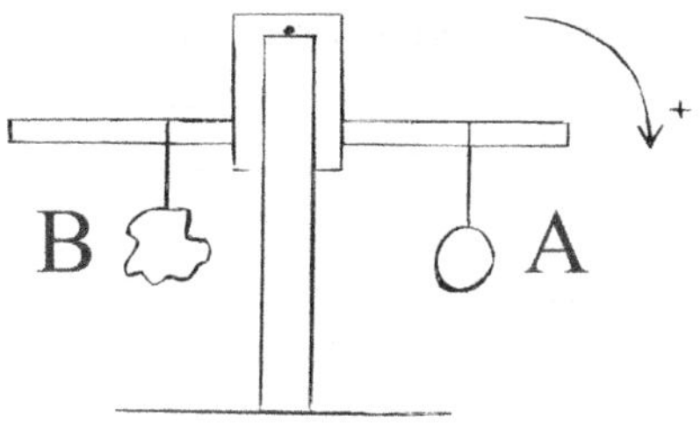

Fig. 9.1. Algebraic torque.

for the case of a lever in equilibrium, but also as a basis for the study of more complex phenomena like the rotational motion of rigid bodies, *etc.*

With this torque concept we can also derive the empirical result that in equilibrium the CG of a rigid body must be along the vertical line passing through the point of suspension. To this end we need to postulate that the weight of any body behaves as if it were concentrated at its CG, acting downwards. As the force exerted upon the fulcrum or PS does not exert any torque upon the lever (because it acts at zero distance to the support and, therefore, has an arm of zero length), there remains only the torque exerted by the body. And this torque only goes to zero when the PS and the CG are along a vertical line.

We can also deal with the torque algebraically. In this case we define a tendency to rotate in one direction (for instance, the rotation of the lever in the vertical plane lowering body A and raising body B at the other side of the lever) as due to a positive torque. We also define a tendency to rotation in the opposite direction as due to a negative torque. In the case of the Fig. 9.1, for instance, the weight of A would exert a positive torque upon the lever, while the weight of B would exert a negative torque. In this case the fundamental postulate might be expressed as follows: The algebraic sum of all torques acting upon a rigid body must be null in order for the body to remain in equilibrium after release from rest, without rotation around a fixed axis.

If we have N bodies on one side of the lever and M bodies on the other side, the basic postulate can be generalized by the principle of superposition. That is, we postulate that the lever will remain in equilibrium if

$$\sum_{i=1}^{N} \frac{P_i}{P_0} \frac{d_i}{d_0} = \sum_{i=N+1}^{N+M} \frac{P_i}{P_0} \frac{d_i}{d_0} .$$

Here P_i is the weight of body i acting at a horizontal distance d_i from the vertical plane passing through the fulcrum of the lever. In addition, P_0 and d_0 are a weight and a distance chosen arbitrarily. We can choose, for instance, $P_0 = P_1$ and $d_0 = d_1$, *etc.* If one of these sums is bigger than the other, we postulate that the side with the greater sum will move toward the Earth if the lever is released from rest, with the other side moving away from it.

Although this theoretical deduction of the law of the lever beginning with the previous definitions and postulates is correct, it should be emphasized that the concept of torque of a force was suggested historically by empirical knowledge of the law of the lever. That is, it was the experimental fact that two bodies equilibrate one another upon a lever when the ratio of their distances to the fulcrum is inversely proportional to the ratio of their weights which suggested the creation of the torque concept. Suppose, for instance, that nature behaved in such a way that the experimental law of the lever were given by the relation $P_A/P_B = (d_B/d_A)^m$, with $m = 2$ or another value. In this case it would be natural to define another magnitude proportional to $(P_i/P_0)(d_i/d_0)^m$, instead of the usual torque proportional to $(P_i/P_0)(d_i/d_0)$. We could then postulate that the net algebraic value of this new magnitude must go to zero in order to have equilibrium. In this case we could derive the new law of the lever theoretically.

What we want to emphasize is that the traditional definitions of torque and center of gravity (as proportional to the distance between the fulcrum and the point of application of the force), together with the postulate that the algebraic sum of all torques acting upon a body in equilibrium must be zero, are only justifiable because they lead to the correct law observed in nature. These definitions and postulates were suggested by the experimental law. When we discover the limits of validity of any specific law, the relevant concepts and postulates must be modified or generalized in order to adapt to the new experimental knowledge.

9.3 Law of the lever derived from the experimental result that a weight 2*P* acting at a distance *d* from the fulcrum is equivalent to a weight *P* acting at a distance *d* – *x*, together with another weight *P* acting at a distance *d* + *x* from the fulcrum

A very simple way to arrive at the law of the lever utilizes two basic ingredients: (I) Equal weights on opposite sides of the lever equilibrate one another when they act at equal distances from the fulcrum. (II) A weight $2P$ acting at an horizontal distance d to the vertical plane passing through the fulcrum is equivalent to a weight P acting at a distance $d - x$ from the fulcrum, together with another weight P acting at a distance

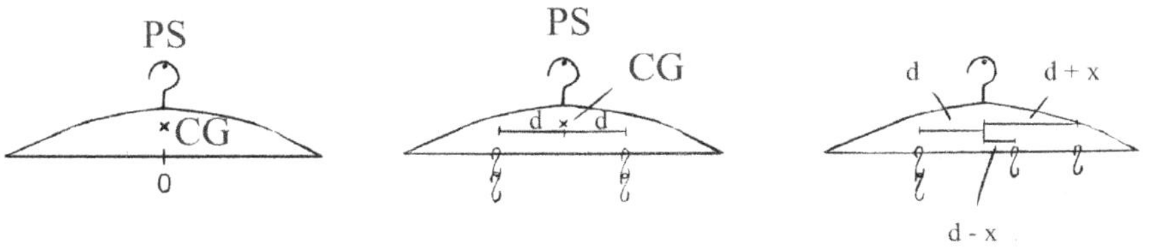

Fig. 9.2. Experimental condition of equilibrium for a lever.

$d+x$ from the fulcrum, as in Fig. 9.2. Here we are utilizing a coat-hanger as a lever. In this case the fulcrum or PS is the horizontal axis passing through the hook of the hanger. We assume that in equilibrium this axis is vertically above the CG of the hanger and above the center 0 of the horizontal section of the hanger. The equivalence mentioned in ingredient (II) refers to the tendency of the lever to rotate around the fulcrum. Ingredient (I) may be considered a definition of equality of weights, while ingredient (II) may be considered an experimental result, or a theoretical postulate. For the moment, we will utilize it as an experimental result. We will treat it as a primitive experimental fact, without trying to explain it.

Ingredient (I), a definition of equality of weights, is represented in the middle of Fig. 9.2. The experimental condition (II) is represented at the right of Fig. 9.2. Experience teaches us that this situation is also a condition of equilibrium.

Assuming condition (II), it is easy to arrive at the law of the lever without imposing any limit upon the possible values of x. To see this we begin with two equal weights P acting at the same distance d on one side of the fulcrum, equilibrated by two other equal weights P acting at the same distance d on the other side. By moving one of the weights on the right hand side to the position $d-x$ and the other weight on the right hand side to the position $d+x$, with $x = 2d$, we end up with the situation shown in Fig. 9.3: *i.e.*, a lever in equilibrium with a weight $3P$ at the distance d from the fulcrum, together with a weight P at the distance $3d$ in the other side of the fulcrum. This is a particular case of the law of the

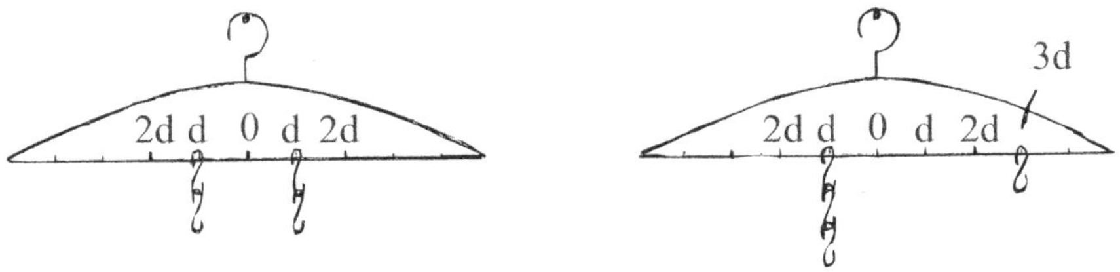

Fig. 9.3. A particular case of the law of the lever for which $P_A/P_B = d_B/d_A = 3$.

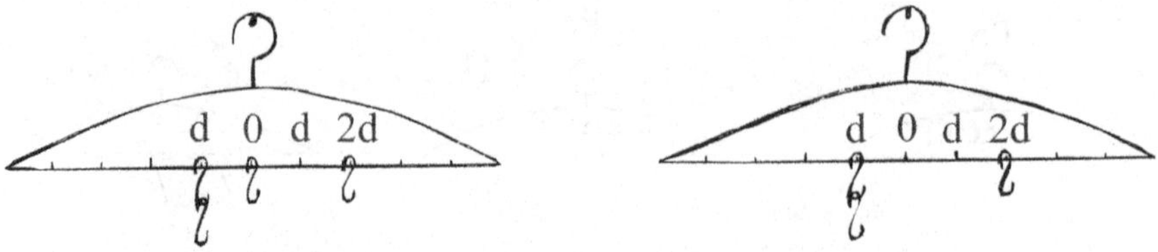

Fig. 9.4. A particular case of the law of the lever for which $P_A/P_B = d_B/d_A = 2$.

lever, since we have $P_A/P_B = d_B/d_A = 3$.

If we had made $x = d$ we would arrive at the equilibrium situation shown on the left side of Fig. 9.4. As one of the weights is along the vertical plane passing through the fulcrum and CG of the lever, it can be removed without affecting the equilibrium. In this case we end up in the situation of equilibrium shown on the right hand side of Fig. 9.4. That is, a lever in equilibrium with a weight $2P$ at the distance d from the fulcrum, together with another weight P at the distance $2d$ on the other side of the fulcrum. And this is another particular case of the law of the lever for which $P_A/P_B = d_B/d_A = 2$.

We now begin with three bodies of equal weight P acting at the same distance d on one side of the fulcrum, equilibrated by three other bodies of equal weight P acting at the same distance d on the other side of the fulcrum. We do not touch the bodies on the left side and consider only the bodies on the right side. We can preserve equilibrium by moving one of these bodies to the right, away from the fulcrum by a distance

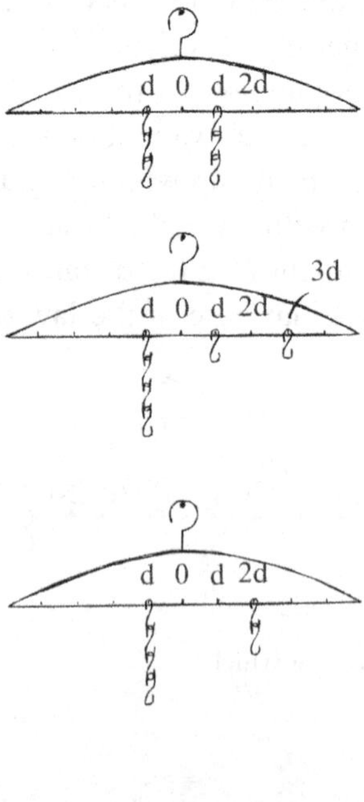

Fig. 9.5. A particular case of the law of the lever for which $P_A/P_B = d_B/d_A = 2$.

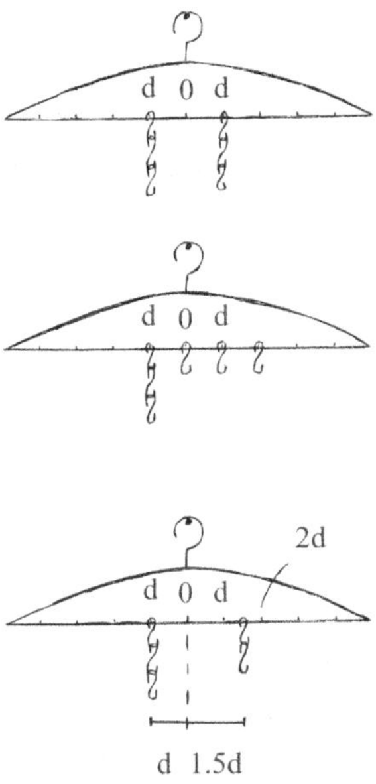

Fig. 9.6. A particular case of the law of the lever for which $P_A/P_B = d_B/d_A = 1.5$.

$x = 2d$, provided that we move one of these bodies to the left by the same distance $2d$ at the same time, while the third body remains fixed in its present position. We end up in the intermediate case shown in Fig. 9.5, *i.e.*, a weight $4P$ at the distance d on one side of the fulcrum, a weight P at the distance d on the other side of the fulcrum, and a weight P at a distance 3d on the same side of the fulcrum. We can preserve equilibrium by joining the latter two bodies at their midpoint, as in the last situation of Fig. 9.5. We then end up with another special case of the law of the lever for which $P_A/P_B = d_B/d_A = 2$. This is the same value obtained before, although this time we did not need to remove a body from the lever.

We now begin once more with three bodies of equal weight P on either side of the lever, acting at a distance d from the fulcrum. By moving one of the bodies on the right hand side to the distance $d - x = 0$ from the fulcrum and another one to the distance $d + x = 2d$ from the fulcrum ($x = d$), we end up in the equilibrium situation shown in Fig. 9.6. As the body which is along the vertical plane passing through the fulcrum and CG of the lever does not disturb the equilibrium, we can remove it from the system. By joining the two weights on the right hand side at their midpoint, we end up with the third case of equilibrium shown in Fig. 9.6, *i.e.*, a weight $3P$ acting at a distance d from the fulcrum and another weight $2P$ acting at a distance $1.5d$ on the other side of the fulcrum. This is another special case of the law of the lever for which $P_A/P_B = d_B/\mathrm{d_A} = 3/2 = 1.5$.

If we had begun with 5 equal clips in either side of the lever, acting at the same distance from the fulcrum, we could have arrived at the same relation without removing a body from the lever.

It is easy to extend this analysis to other cases. This shows how to derive the law of the lever starting from the experimental result that a weight $2P$ acting at an horizontal distance d to the vertical plane passing through the fulcrum is equivalent to a weight P acting at a distance $d-x$ from the fulcrum, together with another weight P acting at a distance $d+x$ from the fulcrum.

9.4 Law of the lever as derived by Duhem utilizing a modification of work attributed to Euclid

The previous procedure seems to be at the origin of one of the oldest theoretical proofs of the law of the lever known to us. This information is taken from Duhem and Clagett.[1] The main idea here is to consider the experimental condition (II) introduced above as a theoretical postulate.

Here we present the main elements of a work of mechanics attributed to Euclid, the famous author of the book *The Elements of Geometry*, who lived in Alexandria around 300 B.C. Although no works on mechanics were attributed to Euclid in Antiquity, many Arabic authors mention works by Euclid on this subject. Three fragments which survived are attributed to him. The titles given to these works are: *Book on the Balance*; *Book on the Heavy and Light*; and *Book on Weights According to the Circumference Described by the Extremities*. What interests us here is the first of these books, which was translated into French in 1851 from its Arabic version (there is no known version of this book in Greek or in Latin). An English translation of this work has been made.[2]

The book begins with a definition and two axioms.[3] Text between square brackets is Clagett's:

> 1. [Definition] Weight is the measure of heaviness and lightness of one thing compared to another by means of a balance.
>
> 2. [Axiom I] When there is a straight beam of uniform thickness, and there are suspended on its extremities two equal weights, and the beam is suspended on an axis at the middle point between the two weights, then the beam will be parallel to the plane of the horizon.

[1] P. Duhem, *The Origins of Statics* (Dordrecht: Kluwer, 1991), Chap. V; M. Clagett, *The Science of Mechanics in the Middle Ages* (Madison: The University of Wisconsin Press, 1979), Chap. I and Document I.1.

[2] Clagett, *The Science of Mechanics*, Document I.1, pp. 24-30.

[3] *Ibid.*, p. 24.

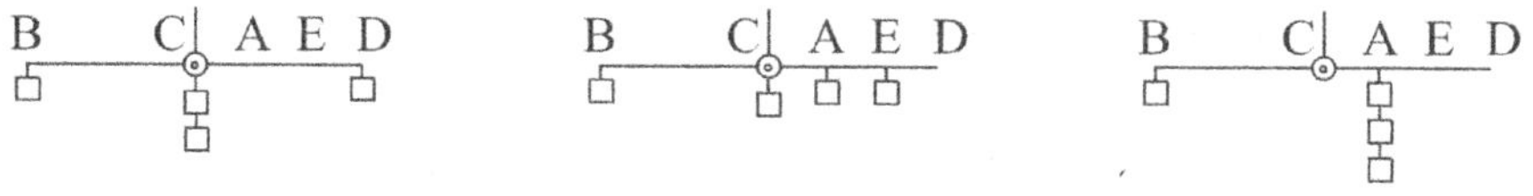

Fig. 9.7. Duhem's proof of the law of the lever.

> 3. [Axiom II] When two weights – either equal or unequal – are placed on the extremities of a beam, and the beam is suspended by an axis on some position of it such that the two weights keep the beam on the plane of the horizon, then if one of the two weights is left in its position on the extremity of the beam and from the other extremity of the beam a straight line is drawn at a right angle to the beam in any direction at all, and the other weight is suspended on any point at all of this line, then the beam will be parallel to the plane of the horizon as before.
>
> This is the reason that the weight is not changed when the cord of one of the two sides of the balance is shortened and that of the other is lengthened.
>
> [Propositions] (...)

The author of this work demonstrates four propositions. The last one contains the law of the lever. In the next section we will discuss this procedure.

For the time being we will follow a modification of this argument which was proposed by Pierre Duhem when he analyzed this work.[4] Duhem postulates two extra axioms, namely (text between square brackets is ours):

> Axiom III. If the weights are maintaining the beam of a balance parallel to the horizon and if one suspends an additional weight to the beam's point of suspension, the beam remains parallel to the horizon.
>
> Axiom IV. If any number of weights maintain the beam of a balance parallel to the horizon, and if Z and D are two of these weights [equal to one another] suspended from the same arm of the beam and if one moves weight Z by a given length away from the point of suspension of the balance and if one moves weight D by the same length towards the point of suspension, then the beam will remain parallel to the horizon.

These axioms lead to an elegant demonstration of the law of the lever. It can be summarized as follows. Let BD be the beam of a lever with C being its fulcrum or point of support, and BC = CD. Suppose four equal bodies of weight P, one suspended at B, another at D, and two at C, as in Fig. 9.7. By axioms I, II and III, the lever will remain in equi-

[4] Duhem, *The Origins of Statics*, pp. 47 to 51.

librium, with its beam at rest horizontally. We divide CD into three equal parts by the points A and E, such that CA = AE = ED = CD/3. We now move one of the bodies which was at C to the point A, while simultaneously moving the body which was at D to the point E, as in the middle of Fig. 9.7. By axiom IV the lever will remain in equilibrium horizontally. By axiom IV it will remain in equilibrium if we move the body that remained in C to the point A, provided we simultaneously move the body that was at E to the point A, as in Fig. 9.7. We then see that the lever in its final configuration of equilibrium will have a weight P at a distance d from the fulcrum and another weight 3P at a distance $d/3$ on the other side of the fulcrum. In other words, we arrive at a particular case of the law of the lever. It is easy to generalize this demonstration.

This demonstration of the law of the lever depends both upon the condition of equilibrium given by equal weights acting at equal distances on opposite sides of the fulcrum, and upon axiom IV. And this is not an obvious axiom. It is justified only because it is in agreement with experimental results. Suppose that nature behaved in such a way that equilibrium resulted when

$$\sum_{i=1}^{N} \frac{P_i}{P_0}\left(\frac{d_i}{d_0}\right)^{\alpha} = \sum_{i=N+1}^{N+M} \frac{P_i}{P_o}\left(\frac{d_i}{d_0}\right)^{\alpha},$$

with $\alpha \neq 1$. In this case axiom IV would no longer hold.

As we will see in the next Section, the original procedure attributed to Euclid contains only the first two axioms. Euclid derived an analogous to Duhem's axiom IV as proposition 2 of his work, based on the first two axioms.

9.5 Proof of the law of the lever by an experimental procedure suggested by a work attributed to Euclid

We present here some experiments which illustrate how to derive the law of the lever in a very interesting way. These experiments were suggested by the *Book on the Balance*, attributed to Euclid.

Up to now we have been dealing with levers composed of horizontal beams which can turn in a vertical plane around a horizontal axis which is orthogonal to the beam. The procedure we will assume here is a different one. We now use a homogeneous rigid rectangle (or square) which remains in equilibrium in a horizontal plane, supported on a vertical stick placed under the center of the rectangle. We place three equal bodies upon this horizontal plane, studying the conditions in which the plane remains in equilibrium. We should attach a sheet of graph paper to the rectangle. This simplifies the analysis as we have now a Cartesian plane above it. We place two orthogonal axes, x and y, parallel to the sides of the rectangle and to the lines of the sheet of paper. We choose the origin of the coordinate system, (0, 0), to be at the center of the rectangle.

Materials: The rectangle can be made of pasteboard and the lines drawn upon it. An alternative procedure is to attach a sheet of graph paper to the pasteboard. The three bodies to be placed upon the rectangle should have the same size and weight. For instance, they could be three equal screw-nuts. During the experiments these nuts will slide over the plane and fall to the ground many times. To prevent this hindrance we should place some glue under the nuts, or attach a thin layer of modeling clay below them, so that they can be attached to any point of the rectangle. Another very interesting alternative is to utilize a metal rectangle (like the picture frame). In this case the three bodies can be small magnets like the ones used to attach the pictures to these frames. The size of the rectangle could be, for instance, 10 cm × 15 cm. The separation between the lines of the graph paper can be 0.5 cm or 1 cm. The vertical stick to be placed below the center of the rectangle can be a bamboo barbecue skewer with its tip stuck in a piece of clay. Any other appropriate stand can be used. The important point is that the upper plane surface of the support (bamboo stick, bottle cap, *etc.*) should not be too small or too large. If it is too small, the equilibrium becomes too unstable and it may be difficult to balance the rectangle on it. If it is too large, it will be very easy to balance the rectangle on it, but it will be difficult to establish the precise conditions which yield the equilibrium of the three equal bodies. As a convenient measure we can use a support such that, when the rectangle remains in equilibrium with the three pieces in adequate positions above it, this equilibrium will be disturbed when a single body moves one or two units of length along the x or y axes, *i.e.*, in such a way that the rectangle falls to the ground with this change of configuration, so that the lack of equilibrium can be easily perceived.

Let us then suppose that we have our graph paper rectangle. The first thing to do is to balance it horizontally, supporting it upon the stick placed under the origin (0, 0) of the rectangle. Next, we equilibrate the rectangle with the three pieces of equal weight placed above it. Let us call them P_1, P_2 and P_3. Initially we put them at $(x, y) = (-5, 0)$, $(0, 0)$ and $(5, 0)$, respectively, as in Fig. 9.8. Because this is a symmetrical configuration around the origin, equilibrium must be established. If this does not happen we must first find the reason before we proceed. It may be due to the fact that the three pieces do not have exactly the same

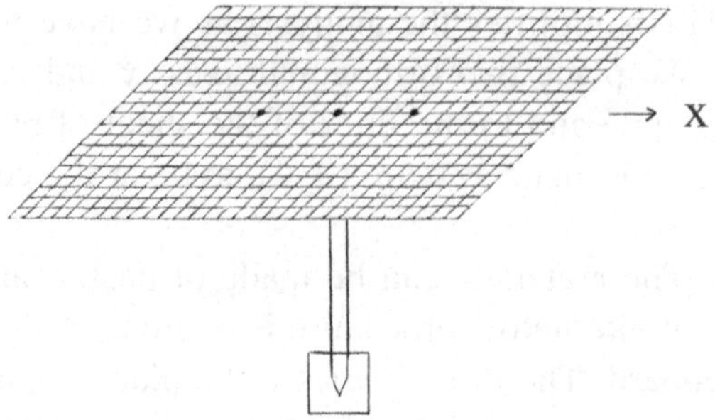

Fig. 9.8. Another procedure to obtain the law of the lever.

weight; or the stick may not be placed exactly below the center of the rectangle; or it is not vertical; or its upper surface is not horizontal, *etc.*

We are now ready to begin the main experiments.

Experiment

We move piece P_2 to the location $(x, y) = (0, 2)$. It should be observed that the system falls to the ground, with this piece moving toward the Earth. On the other hand, when we move P_2 to $(x, y) = (0, 2)$ and P_1 to $(x, y) = (-5, -2)$, leaving P_3 at $(x, y) = (5, 0)$, the system remains in equilibrium horizontally after being released from rest, as in Fig. 9.9.

The result of this experiment can be generalized to other cases. Suppose that we have a set of bodies in equilibrium above a horizontal plane supported by a vertical stick placed below one of its points. We consider the position of this stick the origin of an orthogonal system of coordinates (x, y). If we move one of the bodies from position (x_1, y_1) to position $(x_1 + d, y_1)$ and, simultaneously, move another body of equal weight from position (x_2, y_2) to position $(x_2 - d, y_2)$, the system will remain in equilibrium. It will also remain in equilibrium when we move two pieces of the same weight the same distance in the opposite direction along the y axis, or in any other direction.

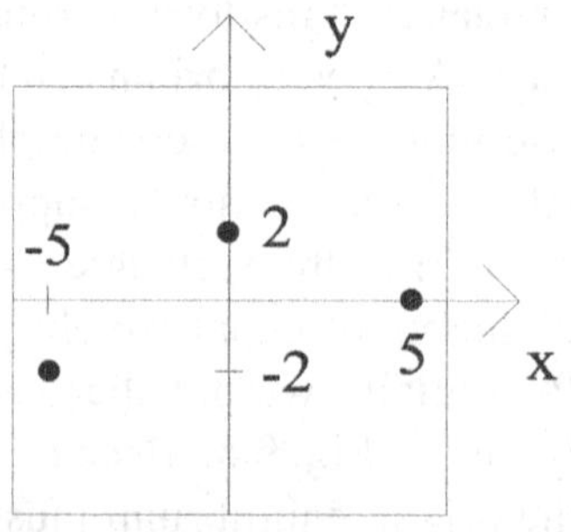

Fig. 9.9. Experimental condition of equilibrium when this rectangle is supported by a vertical stick below the origin.

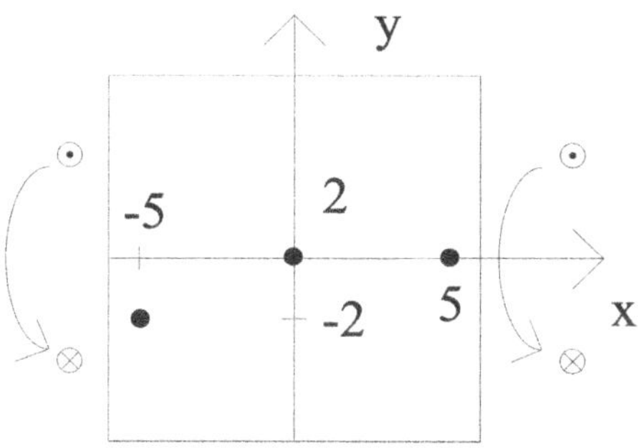

Fig. 9.10. Direction of rotation of the plane.

Experiment

Let us now invert the order of the movements. We again begin with the three bodies P_1, P_2 and P_3 located at $(x, y) = (-5, 0)$, $(0, 0)$ and $(5, 0)$, respectively. We now move only piece P_1 to $(x, y) = (-5, -2)$, holding the rectangle in place with our hands. We now look carefully at the rectangle. When we release the system slowly from rest, what is observed is that the whole side with $y < 0$ tends to fall to the ground, while the side $y > 0$ moves away from the Earth. On the other hand, we perceive no difference between the sides with $x > 0$ and $x < 0$. In other words, none of these sides tends to fall, as indicated by Fig. 9.10. And this is somewhat surprising because body P_1 is not located symmetrically in relation to the origin of the axis x, as it is offset from the vertical extended upwards through the stick.

We can express this finding as follows. Suppose we have a rigid system in equilibrium in a horizontal plane, capable of turning in any direction around a vertical stick, with several bodies on the horizontal plane. If only one of these bodies is moved in a certain direction on the horizontal plane, the system looses equilibrium only in this direction, tending to move closer to the ground, without disturbing the equilibrium in any direction orthogonal to this displacement. For instance, in the previous example the body P_1 moved along the negative y direction. The side of the rectangle characterized by $y < 0$ tended to fall to the ground, while the side $y > 0$ tended to move upwards. This experiment gives empirical support to the second axiom of Euclid presented above.

Experiment

We draw two circles of the same radius on the rectangle in such a way that they touch one another at a single point. If the circles have a radius of 5 units, for instance, the centers of the circles can be located at $(x, y) = (-5, 0)$ and $(5, 0)$. In this case the point of contact is the origin

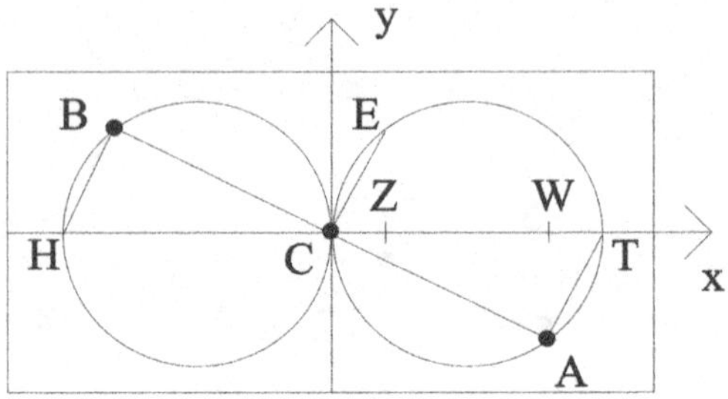

Fig. 9.11. Euclid's procedure to derive the law of the lever.

(0, 0). Let ACB be the straight line passing through the points A = (8, –4), C = (0, 0) and B = (–8, 4). Let H and T be the edges of the circles along the x axis, that is, with H = (–10, 0) and T = (10, 0). We draw three parallel straight lines HB, CE, and AT, with E = (2, 4). The projections of E and of A on the x-axis are called Z and W, respectively, such that Z = (2, 0) and W = (8, 0), as in Fig. 9.11. Experiment shows that this rectangle remains in equilibrium horizontally when supported on a stick placed under its origin. This can be understood by considerations of symmetry.

Experiment

We place three bodies P_1, P_2, and P_3 of equal weight at positions B, C, and A, respectively. Experiment shows that this rectangle remains in equilibrium after being released from rest when supported under its origin (0, 0), as in Fig. 9.11. This can also be understood by considerations of symmetry.

We now move P_1 from B to H and, simultaneously, P_2 from C to E, while keeping P_3 at A. As these displacements were orthogonal to the straight line BCA, were of the same length, in opposite directions and, moreover, as P_1 and P_2 have the same weight, the system remains in equilibrium, based on what we discussed earlier, as in Fig. 9.12.

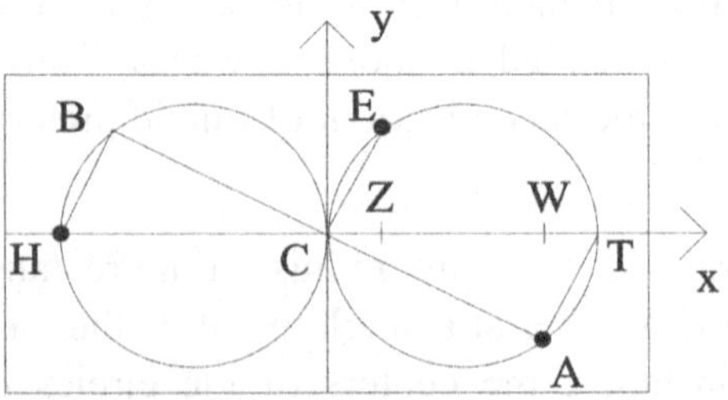

Fig. 9.12. Second step to derive the law of the lever.

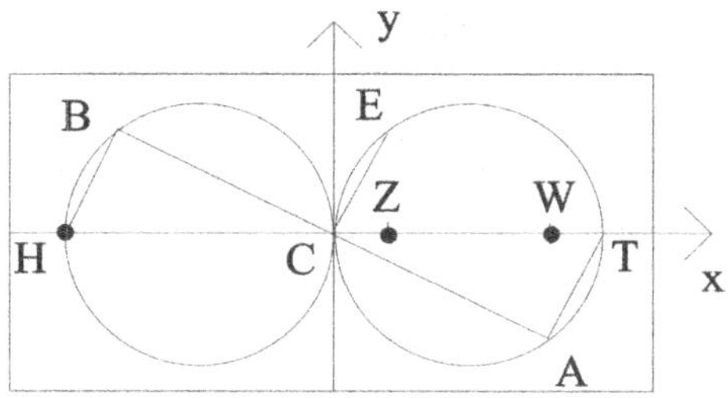

Fig. 9.13. Third step to derive the law of the lever.

Experiment

We now consider the straight line HCT. We begin with the previous equilibrium configuration with the three equal bodies at H, E, and A. We now move P_2 from E to Z, while simultaneously moving P_3 from A to W, keeping P_1 at H. Once more the system remains in equilibrium. After all we have displaced two equal weights by the same amount in opposite directions along the straight line HCT. We end up in the equilibrium configuration shown in Fig. 9.13, in which the three equal weights are located at H = (–10, 0), Z = (2, 0), and W = (8, 0).

By changing the inclination of the straight line BCA to the x-axis and repeating the previous procedure, we will end up with the three equal bodies at the following positions: $P_1 = (-10, 0)$, $P_2 = (a, 0)$, and $P_3 = (10 - a, 0)$, where magnitude a can have any value between 0 and 10. We conclude that a weight at a certain distance d from the origin is balanced by two other equal weights placed on the other side of the fulcrum at the following distances from the origin: a and d – a.

As we saw previously, from this last result it is possible to derive the law of the lever experimentally. A particular example is where $a = d/2$. In this case we arrive at the equilibrium condition of a body P at a distance d from the fulcrum equilibrated by another weight $2P$ at a distance $d/2$ on the other side of the fulcrum, which is a particular case of the law of the lever.

The interesting aspect of this experimental procedure utilizing a rectangle in horizontal equilibrium is that we did not need to begin with this last result. Rather than start from it, we derived this result beginning from the fact that when we displace a body over a system which was originally in equilibrium supported on a vertical stick, the plane looses equilibrium only in this direction. That is, this displacement does not affect the equilibrium of the plane in directions orthogonal to the displacement.

9.6 Theoretical proof of the law of the lever attributed to Euclid

In the theoretical work attributed to Euclid, available in English,[5] the law of the lever is derived from the theoretical postulate of the previous experimental result. This is the essence of the second postulate presented earlier, namely:

> 2. [Axiom I] When there is a straight beam of uniform thickness, and there are suspended on its extremities two equal weights, and the beam is suspended on an axis at the middle point between the two weights, then the beam will be parallel to the plane of the horizon.
>
> 3. [Axiom II] When two weights—either equal or unequal—are placed on the extremities of a beam, and the beam is suspended by an axis on some position of it such that the two weights keep the beam on the plane of the horizon, then if one of the two weights is left in its position on the extremity of the beam and from the other extremity of the beam a straight line is drawn at a right angle to the beam in any direction at all, and the other weight is suspended on any point at all of this line, then the beam will be parallel to the plane of the horizon as before.
>
> This is the reason that the weight is not changed when the cord of one of the two sides of the balance is shortened and that of the other is lengthened.
>
> [Propositions] (...)

The main aspect of this second axiom is the postulate that the equilibrium of a horizontal beam is not disturbed when a body moves orthogonally to this beam "in any direction at all." Suppose the beam is along the x-axis in horizontal equilibrium, supported by a vertical stick placed under one of its points. In this case a body suspended on the beam can move a distance d in the vertical z direction, or along the horizontal y axis, or in any direction in the yz-plane, without disturbing the equilibrium of the beam along the x-axis. That is, displacement of the body along the yz-plane will not cause the side $x > 0$ of the beam to move upward or downward, the same being true for the side $x < 0$.

These two postulates are presented as follows in the English translation of Duhem's book:[6]

> Axiom I. When two equal weights are suspended from the two extremities of a straight beam of uniform thickness which, in turn, is suspended at the midpoint between the two weights, the beam remains parallel to the plane of the horizon.

[5] Clagett, *The Science of Mechanics*, Document I.I, pp. 24-30.

[6] Duhem, *The Origins of Statics*, p. 50.

> Axiom II. When two equal or unequal weights are attached to the two extremities of straight beam which at one of its points is suspended from a fulcrum so that the two weights maintain the beam parallel to the horizon, and if then, we leave one weight in its place at one extremity and draw a straight line from the other extremity of the beam which forms a right angle to the beam on either side of the beam and if one suspends the other weight from any point at all on this line, the beam will remain parallel to the plane of the horizon. This is the reason why the weight does not change if one shortens the strings of one of the two scale pans or lengthens the strings of the other.

The crucial words "in any direction at all" are replaced in this translation by "on either side of the beam."

As it is an axiom, we cannot derive it from other axioms. We simply postulate it, without explaining the axiom. And we derive consequences from it.

This theoretical axiom can be visualized by the experiments presented earlier. From this axiom we can derive theoretically that a weight P at the position $x = -d$ on one side of the fulcrum, located at $x = 0$, is equilibrated by two other equal weights placed at positions $x = a$ and $x = d - a$. And from this result we can derive the law of the lever, as is done in the work attributed to Euclid.

9.7 Archimedes's proof of the law of the lever and calculation of the center of gravity of a triangle

9.7.1 Law of the lever

Archimedes presented a theoretical deduction of the law of the lever in his works *On the Equilibrium of Planes*, and *The Centres of Gravity of Planes*. It is available in English.[7]

What has reached us seems to have been only a part of a larger work. His proof of the law of the lever is based upon the concept of the center of gravity, which does not appear explicitly defined in any of his works now extant. But from what we have seen in the quotations of Heron, Pappus and Simplicius, who had access to works of Archimedes that are no longer extant, he seems to have defined the CG along the lines of CG8, namely: The center of gravity of any rigid body is a point such that, if the body were to be suspended from that point, released from rest and free to rotate in all directions around this point, the body

[7] T.L. Heath, ed., *The Works of Archimedes* (New York: Dover, 2002), pp. 189-220; E.J. Dijksterhuis, *Archimedes* (Princeton: Princeton University Press, 1987), pp. 286-313 and 346-360.

so suspended will remain at rest and preserve its original position, no matter what its initial orientation relative to the ground.

In Proposition 6 of his work *Quadrature of the Parabola*, he mention the following:[8] "Every body, suspended by any point, assumes an equilibrium state when the point of suspension and the center of gravity are on the same vertical line. This has been demonstrated." The proof of this proposition was probably included in his lost work *On Balances or On Levers*. This proposition offers a practical procedure for finding the CG of a body, as we saw in CG6 and CG7.

In order to prove the law of the lever in his work *On the Equilibrium of Planes*, Archimedes begins with seven postulates, namely:[9]

> I postulate the following:
>
> 1. Equal weights at equal distances are in equilibrium, and equal weights at unequal distances are not in equilibrium but incline towards the weight which is at the greater distance.
>
> 2. If, when weights at certain distances are in equilibrium, something be added to one of the weights, they are not in equilibrium but incline towards the weight to which the addition was made.
>
> 3. Similarly, if anything be taken away from one of the weights, they are not in equilibrium but incline towards the weight from which nothing was taken.
>
> 4. When equal and similar plane figures coincide if applied to one another, their centres of gravity similarly coincide.
>
> 5. In figures which are unequal but similar the centres of gravity will be similarly situated. By points similarly situated in relation to similar figures I mean points such that, if straight lines be drawn from them to the equal angles, they make equal angles with the corresponding sides.
>
> 6. If magnitudes at certain distances be in equilibrium, (other) magnitudes equal to them will also be in equilibrium at the same distances.
>
> 7. In any figure whose perimeter is concave in (one and) the same direction the centre of gravity must be within the figure.

The fundamental postulate which allows Archimedes not only to derive the law of the lever, but also to theoretically locate the CG of many two-dimensional (triangles, parallelograms, trapeziums, circles, semi-circles, parabolic segments) and three-dimensional figures (cones, hemispheres, semi-ellipsoids, paraboloids of revolution, hyperboloids of revolution), is his sixth postulate.

[8] C. Mugler, *Les Oeuvres d'Archimède* (Paris: Budé, 1971). Vol. 2, p. 171; Duhem, *The Origins of Statics*, p. 463.

[9] Heath, ed., *The Works of Archimedes*, pp. 189-190.

The meaning of this crucial postulate has been clarified by Vailati, Toeplitz, Stein and Dijksterhuis.[10] By "magnitudes equal to other magnitudes," Archimedes wishes to say "magnitudes of the same weight." And by "magnitudes at certain distances," Archimedes wishes to say "the centers of gravity of the magnitudes are at the same distances from the fulcrum." Suppose a system of bodies keeps a balance in equilibrium. According to this postulate, Archimedes can replace a certain body A suspended by the beam through its center of gravity located at a horizontal distance d from the vertical plane passing through the fulcrum, with another body B which has the same weight as A, without disturbing equilibrium, provided it is also suspended by the beam at its CG which is at the same horizontal distance d from the vertical plane passing through the fulcrum. Instead of bodies A and B, we can also think of a set of N bodies A_i and M bodies B_i. Equilibrium will not be disturbed when we replace the N bodies A_i with the M bodies B_i, if the total weight of the two sets is the same and if the CG of the set of M bodies B_i acts at the same distance from the fulcrum as the CG of the N bodies A_i.

The algebraic statement of this theorem was given after the mathematical definition CG9. Utilizing this mathematical definition, we can derive this result. In this case it would no longer be a postulate. This theorem can also be stated as: If a body is composed of two or more parts whose centers of gravity are known, then the center of gravity of the composite body can be calculated by regarding its component parts as single particles located at their respective centers of gravity.

A particular example of this postulate is the replacement of a body of weight P located at the distance d from the vertical plane passing through the fulcrum of a lever in equilibrium by a set of two other bodies, namely: a weight $P/2$ located at the distance $d+x$ from the vertical plane passing through the fulcrum, and another weight $P/2$ located at the distance $d-x$ from the vertical plane passing through the fulcrum. In this case the two systems have the same total weight P. Moreover, the centers of gravity of the two systems are located at the same distance d from the vertical plane passing through the fulcrum. In the case of the second system composed of two weights $P/2$, this was proved by Archimedes in the fourth proposition of this work: "If two equal weights have not the same centre of gravity, the centre of gravity of both taken together is at the middle point of the line joining their centres of gravity."[11] From this particular example we can arrive at the law of the lever,

[10] Dijksterhuis, *Archimedes*, pp. 289-304 and 321-322.

[11] Heath, ed., *The Works of Archimedes*, p. 191.

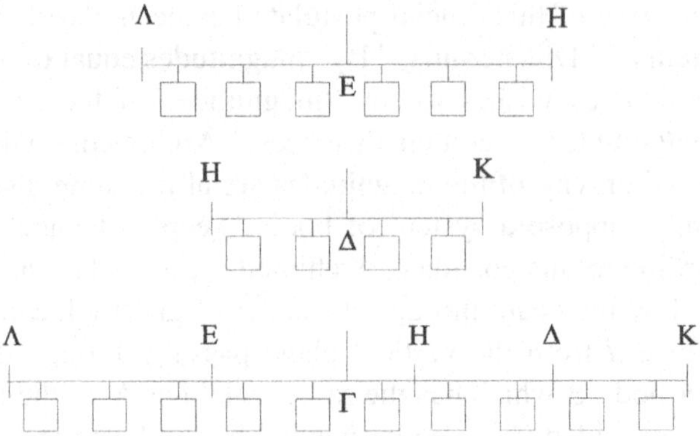

Fig. 9.14. Archimedes's procedure for deriving the law of the lever.

as we saw earlier in the procedure attributed to Euclid. Archimedes presented a general proof of the law of the lever which is valid for commensurable magnitudes as well as incommensurable magnitudes.

The advantage of the postulate due to Archimedes as compared with the analogous postulate from Euclid is the generality implied by Archimedes's approach. Using this postulate, he derived both the law of the lever and the correct calculation of the CG of all one-, two-, and three-dimensional figures, as discussed above.

We now present the main elements of his proof of the law of the lever. He considers three situations, namely: (A) A set of $2N_1$ magnitudes of the same weight P suspended by their centers of gravity along a rectilinear lever, with these magnitudes evenly spaced. We present a concrete example with $N_1 = 3$ and with the spacing between adjacent magnitudes given by the length w. The CG of this system of magnitudes is the point E, which is at the midpoint of these magnitudes, as in the first situation in Fig. 9.14. (B) A system of $2N_2$ magnitudes of the same weight P suspended by their centers of gravity along a rectilinear lever, with these magnitudes evenly spaced. We present a concrete example with $N_2 = 2$ and with the spacing between adjacent magnitudes given by w. The CG of this system of magnitudes is the point Δ, which is at the midpoint of these magnitudes, as in the second situation in Fig. 9.14. (C) A system of $2N_1 + 2N_2$ magnitudes of the same weight P suspended by their centers of gravity along a rectilinear lever, with these magnitudes evenly spaced. We present a concrete example with $N_1 = 3$, $N_2 = 2$ and with the spacing between adjacent magnitudes given by w. The CG of this system of magnitudes is the point Γ, which is at the midpoint between these magnitudes, as in the last situation in Fig. 9.14.

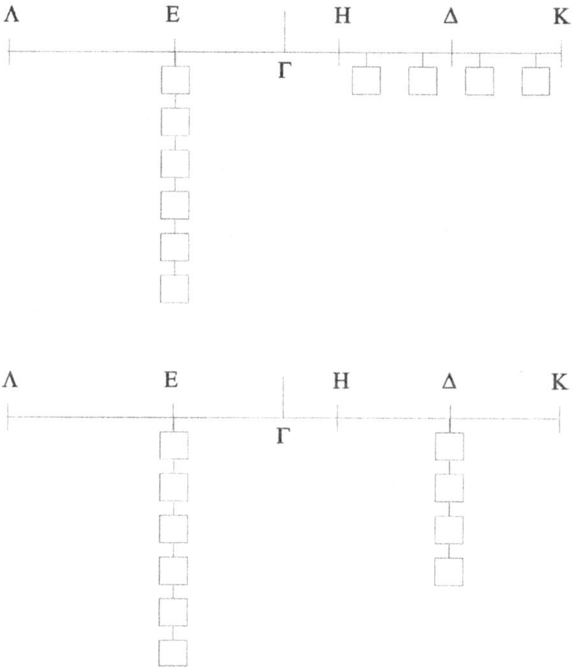

Fig. 9.15. Second and third steps in Archimedes's procedure for deriving the law of the lever.

That the CG of these three situations are located at the points E, Δ and Γ, respectively, was proven by Archimedes in Corollary II of Proposition 5 of his work: "If there be an even number of magnitudes with their centres of gravity situated at equal distances on one straight line, and if the two middle ones be equal, while those which are equidistant from them (on each side) are equal respectively, the centre of gravity of the system is the middle point of the line joining the centres of gravity of the two middle ones."[12] But this fact does not depend upon the linear aspect of the usual law of the lever. That is, this result can be proved by considerations of symmetry, even if the law of the lever were given by

$$\sum_{i=1}^{N} \frac{P_i}{P_0}\left(\frac{d_i}{d_0}\right)^{\alpha} = \sum_{i=N+1}^{N+M} \frac{P_i}{P_0}\left(\frac{d_i}{d_0}\right)^{\alpha},$$

with $\alpha = 1$ or with $\alpha \neq 1$. That is, suppose that it is found experimentally that a lever only remained in equilibrium when the previous relation was valid with a specific value of α. Even if this were the case, the CG of the three situations presented earlier would be located at the points E, Δ, and Γ, regardless of the value of α.

[12] *Ibid.*

But now we must appeal to the crucial postulate 6. It stipulates that in the third situation already presented the set of $2N_1$ bodies can be replaced by a single body A of weight $P_A = 2N_1P$ acting at point E, as shown in the first situation of Fig. 9.15, which we will call situation (D). That is, if situation (C) was an equilibrium situation, then postulate 6 guarantees that situation (D) will also be an equilibrium situation. It is also possible to replace the set of $2N_2$ bodies with a single body B of weight $P_B = 2N_2P$ acting at point Δ. This is shown in the second situation of Fig. 9.15, which we will call situation (E). That is, postulate 6 guarantees that situations (D) and (E) will be equilibrium situations, as was the case with situation (C).

This is the law of the lever, because the weight P_A is to the weight P_B as the distance ΔΓ is to the distance ΕΓ.

Suppose now that nature behaved in such a way that the law of the lever were of the type

$$\sum_{i=1}^{N} \frac{P_i}{P_0}\left(\frac{d_i}{d_0}\right)^{\alpha} = \sum_{i=N+1}^{N+M} \frac{P_i}{P_0}\left(\frac{d_i}{d_0}\right)^{\alpha},$$

with $\alpha \neq 1$. That is, suppose that a lever only remained in equilibrium when this equation was satisfied. In this case situation (C) would still be an equilibrium configuration. But when we tried to go to situations (D) and (E), equilibrium would be disturbed. These two configurations would not remain in equilibrium when the system was released from rest. This shows that in this hypothetical situation Archimedes's postulate 6 would no longer be valid.

9.7.2 CG of a triangle

We now analyze certain aspects of the calculation of the CG of a triangle given by Archimedes. This CG coincides with the intersection of the medians, which are the straight lines connecting the vertices to the midpoints of the opposite sides. The importance of this result is that it is only valid for a law of the lever which is linear with distance. On the other hand, the CG of a circle or rectangle would still be the geometric centers of these figures even if the law of the lever were quadratic or cubic in distance, as can be seen by arguments of symmetry. This means that the calculation of the CG of a triangle is the first non trivial result for the CG of a two-dimensional figure.

Archimedes considers a generic scalene triangle ΑΒΓ. In Proposition 13 he then shows that the CG must be along the straight line connecting any vertex to the midpoint of the opposite side: "In any triangle the centre of gravity lies on the straight line joining any angle to the

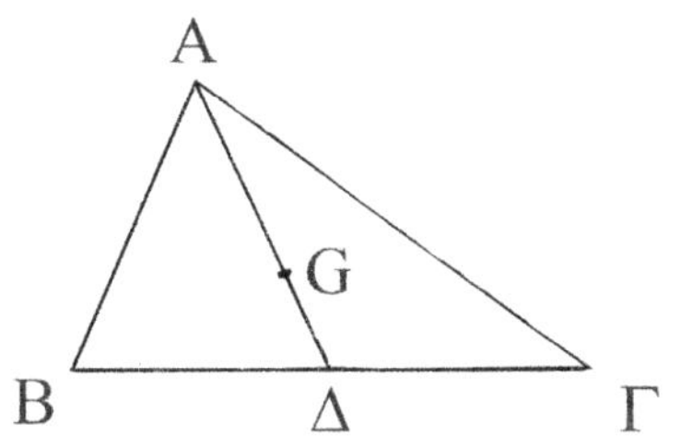

Fig. 9.16. Center of gravity of a triangle.

middle point of the opposite side."[13] In Proposition 14 he concludes that "the centre of gravity of any triangle is at the intersection of the lines drawn from any two angles to the middle points of the opposite sides respectively."[14] If Δ is the midpoint of the side BΓ in the next triangle, this means that the CG must be at some point G along the straight line AΔ, as in Fig. 9.16. Archimedes presents two proofs of this fact. The two proofs suppose that the CG is not along this line AΔ, which leads to a logical contradiction. This means that the CG must be along the line AΔ, which is what he wanted to prove.

Here we explore the opposite viewpoint. We begin supposing that the CG of the triangle is at some point G along the line AΔ connecting the vertex to the midpoint of the opposite side. We then show that we do not arrive at any logical contradiction with this reasoning. Moreover, we arrive at the ratio between AG and GΔ. With this simplified analysis we can make comprehension of Archimedes's original proof easier. We present all the postulates which used in the proof explicitly.

From Postulate 7 we conclude that the CG must be inside the triangle ABΓ. We then suppose that it is at a point G along the straight line AΔ, where Δ is the midpoint of the side BΓ. Let E be the midpoint of the side AB and Z the midpoint of the side AΓ, as in Fig. 9.17. We join the segments EΔ, ZΔ, and EZ. The segment EΔ is parallel to the side AΓ; the segment EZ is parallel to the side BΓ; and the segment ΔZ is parallel to the side AB. This leads to the result that $B\Delta = \Delta\Gamma = EZ = B\Gamma/2$, $BE = EA = \Delta Z = BA/2$, $AZ = Z\Gamma = E\Delta = A\Gamma/2$. We then obtain four

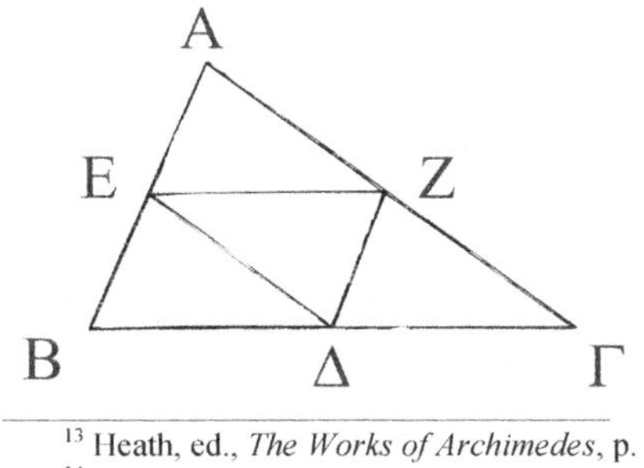

Fig. 9.17. Step in finding the CG of a triangle.

[13] Heath, ed., *The Works of Archimedes*, p. 198.
[14] *Ibid*., p. 201.

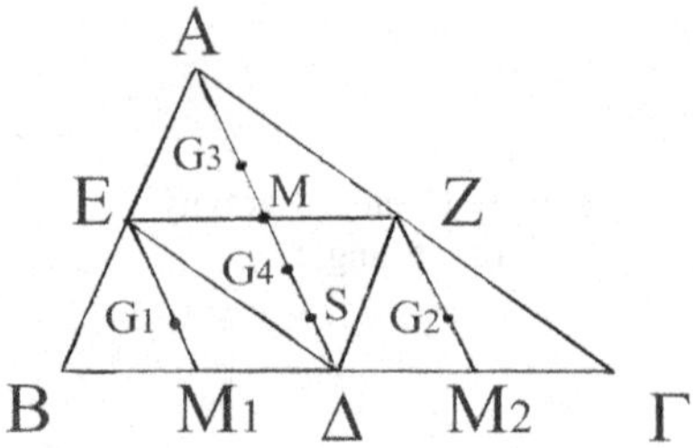

Fig. 9.18. Final step in finding the CG of a triangle.

equal triangles: EBΔ, ZΔΓ, AEZ, and ΔZE, as in Fig. 9.17. These four triangles are similar to the original triangle ABΓ. The area and weight P of each one of them are equal to a quarter of the area and weight of the original triangle $P_{EB\Delta} = P_{Z\Delta\Gamma} = P_{AEZ} = P_{\Delta ZE} = P_{AB\Gamma}/4$.

Let M be the midpoint of the segment EZ, which is also the midpoint of the segment AΔ. Let M_1 be the midpoint of the segment BΔ and M_2 be the midpoint of the segment ΔΓ. We join EM_1, ZM_2 and AΔ. By postulate 5 the centers of gravity of the triangles EBΔ, ZΔΓ, AEZ and ΔZE will be at the points G_1, G_2, G_3, and G_4 along the straight segments EM_1, ZM_2, AM, and ΔM, respectively, situated in such a way that $EG_1 = ZG_2 = AG_3 = \Delta G_4 = AG/2$, as in Fig. 9.18.

By postulates 1 and 6 we conclude that if the original triangle ABΓ was in equilibrium when supported by point G, then it will remain in equilibrium supported by G when we replace the two triangles EBΔ and ZΔΓ by a single body of weight equal to the sum of the weight of these two smaller triangles, acting at the midpoint of the straight line G_1G_2. Let S be this midpoint, located along AΔ. As a matter of fact, in Proposition 4 of this work Archimedes proves the following result: "If two equal weights have not the same centre of gravity, the centre of gravity of both taken together is at the middle point of the line joining their centres of gravity."[15]

By the same token, we can replace the two triangles AEZ and ΔZE with a single body of weight equal to the sum of the weights of these two smaller triangles, acting at the midpoint of the segment G_3G_4, *i.e.*, at the point M. This means that the system will remain in equilibrium supported by G after this replacement.

We then have only two equal weights acting at M and S. Once again we can replace these two weights with a single body having the total weight of the original triangle, acting at the midpoint of the segment MS. And this midpoint of MS must be the CG of the original triangle, the point G. By postulate 5 we have $S\Delta = G_1M_1 = G_2M_2 = G_3M = G_4M = G\Delta/2$. As G is the midpoint of the segment MS, we have

[15] Heath, ed., *The Works of Archimedes*, p. 191.

$G\Delta = (M\Delta + S\Delta)/2$. Combining the last two equalities, we obtain: $G\Delta = (M\Delta + G\Delta/2)/2$. As a result, $2G\Delta - G\Delta/2 = 3G\Delta/2 = M\Delta$. Since $M\Delta = A\Delta/2$, we obtain finally $A\Delta = 3G\Delta$. Since $A\Delta = AG + G\Delta$ we also find that $AG = 2G\Delta$.

We can then conclude that the supposition that the CG of the triangle is along the straight line joining each vertex to the midpoint of the opposite side is a coherent supposition. Moreover, this procedure shows that the CG given by the point G will divide this straight line $A\Delta$ in such a way that $AG = 2G\Delta$.

On the other hand, as $\Delta G_4 = AG/2$, we deduce from the last result that $\Delta G_4 = (2G\Delta)/2 = \Delta G$. In other words, the CG of the triangle ΔZE, which is located at the point G_4, coincides with the CG of the original triangle $AB\Gamma$, which is located at the point G.

We consider these achievements of Archimedes to be some of the greatest scientific accomplishments humankind has produced.

Bibliography

Clagett, M. *The Science of Mechanics in the Middle Ages*. Madison: The University of Wisconsin Press, 1979.

Dijksterhuis, E.J. *Archimedes*. Princeton: Princeton University Press, 1987.

Duhem, P. *The Origins of Statics*. Dordrecht: Kluwer, 1991.

Ferreira, N. *Mecânica*. Projeto RIPE – Rede de Instrumentação para o Ensino. São Paulo: Instituto de Física, USP.

Ferreira, N. *Equilíbrio*. São Paulo: Instituto de Física, USP, 2007. http://www.ludoteca.if.usp.br/

Gaspar, A. *Experiências de Ciências para o Ensino Fundamental*. São Paulo: Ática, 2003.

Heath, T.L. *A History of Greek Mathematics*. Vol. II: From Aristarchus to Diophantus Oxford: Clarendon Press, 1921.

Heath, T.L., ed. *The Works of Archimedes*. New York: Dover, 2002.

Heiberg, J.L. *Archimedis Opera*. Second edition, in 3 volumes. Leipzig: Teubner, 1910-1915.

Héron d'Alexandrie. *Les Mécaniques, ou, L'Élévateur des Corps Lourds*. Paris: Les Belles Lettres, 1988.

Knorr, W.R. "Archimedes and the Elements: Proposal for a Revised Chronological Ordering of the Archimedean Corpus." *Archive for the History of Exact Sciences* **19** (1978/79): 211-290.

Mach, E. *The Science of Mechanics*. La Salle: Open Court, 1960.

Mugler, C. *Les Oeuvres d'Archimède. Paris*: Budé, 1970. Vol. 1: De la Sphère et du Cylindre, La Mesure du Cercle, Sur les Conoïdes et les Sphéroïdes.

Mugler, C. *Les Oeuvres d'Archimède*. Paris: Budé, 1971. Vol. 2: Des Spirales, De l'Équilibre des Figures Planes, L'Arénaire, La Quadrature de la Parabole.

Mugler, C. *Les Oeuvres d'Archimède*. Paris: Budé, 1971. Vol. 3: Des Corps Flottants, Stomachion, La Méthode, Le Livre des Lemmes, Le Problème des Bœufs.

Mugler, C. *Les Oeuvres d'Archimède*. Paris: Budé, 1972. Vol. 4: Commentaires d'Eutocius et Fragments.

Pappus d'Alexandrie. *La Collection Mathématique*. Paris: Blanchard, 1982.

Plutarch. *Marcellus*. http://classics.mit.edu/Plutarch/marcellu.html, translated by John Dryden, site available in 2007.

Rorres C. (organizer), *Archimedes*. http://www.math.nyu.edu/~crorres/Archimedes/contents.html. Site available in 2007.

Symon, K.R. *Mechanics*. Reading: Addison-Wesley, 1971, third edition.

Thomson, G.D. *The Origin of the Zodiac*. Available in 2007 at: http://members.optusnet.com.au/~gtosiris/page9a.html.

Made in the USA
Monee, IL
04 November 2026

67378682R10115